Mercury and Methylmercury Contamination of Terrestrial and Aquatic Ecosystems

Mercury and Methylmercury Contamination of Terrestrial and Aquatic Ecosystems

Editors

Stéphane Guédron
Dario Acha

MDPI • Basel • Beijing • Wuhan • Barcelona • Belgrade • Manchester • Tokyo • Cluj • Tianjin

Editors
Stéphane Guédron
University Grenoble-Alpes,
Univ. Savoie Mont Blanc, CNRS, IRD,
IFSTTAR, ISTerre
France

Dario Acha
Unidad de Calidad Ambiental (UCA)—Instituto de
Ecología—Universidad Mayor de San Andrés,
Campus Universitario de Cota Cota
Bolivia

Editorial Office
MDPI
St. Alban-Anlage 66
4052 Basel, Switzerland

This is a reprint of articles from the Special Issue published online in the open access journal *Applied Sciences* (ISSN 2076-3417) (available at: http://www.mdpi.com).

For citation purposes, cite each article independently as indicated on the article page online and as indicated below:

LastName, A.A.; LastName, B.B.; LastName, C.C. Article Title. *Journal Name* **Year**, *Volume Number*, Page Range.

ISBN 978-3-0365-2153-4 (Hbk)
ISBN 978-3-0365-2154-1 (PDF)

Cover image courtesy of Stéphane Guédron

Contents

About the Editors

Stéphane Guédron has been a researcher in geochemistry and paleoenvironmental studies at the Institute of Research for Development (IRD) within the Institute of Earth Sciences (ISTerre) at University Grenoble-Alpes, France, since 2011. In 2008, he obtained his PhD degree in environmental geochemistry at Grenoble University, France. Within his research, he has studied topics in the fields of the bio-geochemistry of metallic contaminants in terrestrial and aquatic ecosystems, stable isotopes, and paleoenvironmental reconstructions using sediment archives. In addition to this, he has authored 51 peer review publications, and regularly supervises PhD, MSc and undergraduate students, as well as engineer staff in the field and laboratory.

Dario Acha, Ph.D. has been a professor at the Institute of Ecology (IE)—Biology Department, Universidad Mayor de San Andrés (UMSA) since 2010. He obtained his Ph.D. in Environmental and Life Sciences from Trent University, Ontario, Canada. His research predominantly focuses on understanding the role that microorganisms and biota in general play in the biochemical cycle of contaminants (Hg, S, As, C, N, and P) and natural elements in aquatic environments. He also aims to provide research that contributes to the understanding of how global change influences biogeochemical cycles. Recently, his work is slowly moving towards bioremediation, and its potential role in solving metal pollution and eutrophication. He has previously worked in boreal lakes and tropical lowlands; however, his current research surrounds the tropical highlands and endorheic systems. He has co-authored 25 peer-reviewed publications, and supervises M.Sc. and undergraduate theses.

applied sciences

Editorial

Mercury and Methylmercury Contamination of Terrestrial and Aquatic Ecosystems

Stéphane Guédron [1,*] and Dario Acha [2,*]

1 Geochemistry Department, Institut des Sciences de la Terre (ISTerre), University Grenoble Alpes, University Savoie Mont Blanc, CNRS, IRD, IFSTTAR, 38000 Grenoble, France
2 Unidad de Calidad Ambiental (UCA)—Instituto de Ecología—Universidad Mayor de San Andrés, Campus Universitario de Cota Cota, Casilla, La Paz 3161, Bolivia
* Correspondence: stephane.guedron@ird.fr (S.G.); darioacha@yahoo.ca (D.A.)

In 2017, 128 countries signed the Minamata Convention on Mercury (Hg) to protect human health and the environment from the adverse effects of mercury. Although Hg naturally occurs at a global scale, anthropogenic releases of Hg affect its natural biogeochemical cycle. Since the industrial revolution, anthropogenic pressure and Hg contamination have risen, affecting natural ecosystems (e.g., lake eutrophication, deforestation, mining, and climate change) and urban areas (wastewaters, atmospheric emissions, and industrial discharges). Consequently, it has resulted in a general increase in Hg levels in all components of ecosystems, particularly in Hg bioaccumulation in trophic chains, mostly as methylmercury (MeHg). However, Hg and MeHg impacts are dependent on a large variety of abiotic and biotic factors with complex interactions among them, which often results in locally specific responses that we are still trying to understand.

Identifying sources, magnitude, pathways, and mechanisms of Hg and MeHg contamination in terrestrial and aquatic environments is crucial to determine and predict future impacts of this contaminant in the environment. It is also difficult to predict the risks and the Hg toxicity for wildlife and humans. The identification of processes driving Hg release from current and historical sources, key transformations (i.e., methylation/demethylation), and Hg accumulation in terrestrial and aquatic ecosystem is still challenging to assess. These interconnected processes can affect both natural (temperate and extreme) and anthropogenic (urban, agricultural, and mining) areas at various spatial and temporal scales.

Nine articles were accepted following the October 2019 call for a Special Issue (SI) on "Mercury and Methylmercury Contamination of Terrestrial and Aquatic Ecosystems", announced at the 14th International Conference on Mercury as a Global Pollutant—Krakow, Poland (8–13 September 2019). They present a diverse but coherent collection that endeavors to identify the historical evolution of Hg and MeHg levels in aquatic environments and to evaluate the impact of human activities, in particular mining, on receptor ecosystems and food chains.

The need to evaluate the evolution of mercury and methylmercury contamination from historical to modern times is a key to assessing the impact of global regulations and to predicting future trends. Various approaches are presented in this SI, including fossils or modern organisms as bioindicators of trophic contamination. Using Hg levels in the hair of prehistorical animals, Eyrikh et al. [1] showed the first evidence of natural changes in environmental Hg level at the paleoscale (45 to 10 ka yr BP) suggested to be driven by high dust loads in cold periods and thawing permafrost in warm climatic periods. Cossa and Tabard [2] used marine mussels (*Mytilus* spp.) as sentinel organisms to monitor Hg contamination along the Canadian marine coast for the past 40 years. They show that, despite a decrease in Hg concentrations in the Atlantic Ocean's marine waters during the last decades, the difference in two mussels watch surveys between 2016–2019 and 1977–1979 were not significant. Consequently, they suggest rivers as the main Hg

Citation: Guédron, S.; Acha, D. Mercury and Methylmercury Contamination of Terrestrial and Aquatic Ecosystems. *Appl. Sci.* **2021**, *11*, 4807. https://doi.org/10.3390/app11114807

Received: 14 April 2021
Accepted: 17 May 2021
Published: 24 May 2021

source in such estuarine systems. In contrast, in the United States and Canadian Lakes, Brigham et al. [3] showed that regional- to continental-scale decreases in both mercury and sulfate emissions have benefitted aquatic resources, even in the face of global increases in mercury emissions. In long-term biomonitoring of three lakes, the authors found a decline in aqueous MeHg, whereas biotic THg concentrations declined significantly in only one lake. Overall, these studies also show the complexity and mosaic-like characteristics of Hg pollution at the regional to the global scales, underling the need for further local research and tools to understand the global picture better.

Simultaneously monitoring MeHg pollution and its potential toxicity to biota also remains a significant challenge. Evaluating the stress of local biota related to MeHg requires a complex analysis of bioavailability, bioaccumulation, biomagnification, and physiological characteristics of organisms. Even doing this makes it difficult to establish a clear relation between MeHg concentrations and any negative effect on biota. Furthermore, it is challenging to assess MeHg toxicity in the frequent presence of other co-occurring stressors with the potential for synergic or antagonist effects. Ecotoxicological studies offer feasible alternatives to address this complex issue. A good example is how the toxicity of methylmercury to loggerhead turtles was evaluated by Hernández-Fernández et al. [4] using the activity of the enzymes superoxide dismutase (SOD), glutathione S-transferase (GST), and lipid peroxidation by malondialdehyde (MDA) as biomarkers of environmental pollution and oxidative stress.

Another major issue in the anthropogenic perturbation of the Hg biogeochemical cycle is assessing the impact of historical and current gold mining activities. Historical mining activities are still sources of Hg for downstream ecosystems. Samaniego et al. [5] show evidence of the persistent contamination of abandoned and inactive mines in the Philippines, even 50 to 70 years later, releasing high amounts of Hg into riverine and marine environments. In current alluvial gold mines of the Amazon region, Guzmán-Uria et al. [6] highlight the impact of soil degradation on the Hg release into downstream hydrosystems. Hellal et al. [7] demonstrate that particulate Hg and MeHg transport rises downstream from gold mine sites, with the highest emissions during high water regimes (rainy season). In contrast, the highest releases of dissolved MeHg (up to 30% of THg) occur mainly during low water regimes. In downstream receptacle aquatic ecosystems, the methylation of accumulated Hg is generally enhanced. Guédron et al. [8] show that, in mine-impacted high-altitude aquatic ecosystems of the Andes, the production and release of MeHg is driven by sediment effluxes to the water column and diel redox oscillations. Finally, they highlight that, although MMHg loading from the PW to the SW is large, MMHg photodegradation and demethylation by microorganisms control net MMHg accumulation in the water column.

These studies converge toward the idea that, even though Hg is banned, current gold-mining practices still release natural Hg that, added to the Hg inherited from older and modern mining practices, impact downstream ecosystems. The improvement of mining practices is thus key to regulating and mitigating these Hg emissions. Indeed, Couic et al. [9] demonstrate that recent rehabilitation efforts in French Guyana's mining sites show promising results that depend on the type of ecological rehabilitation protocols. They demonstrate that specific plant species (e.g., fabaceous species) positively affect soil quality that can significantly reduce mercury mobility and toxicity.

References

1. Eyrikh, S.; Boeskorov, G.; Serykh, T.; Shchelchkova, M.; Papina, T. Mercury in Hair of Mammoth and Other Prehistorical Mammals as a Proxy of Hg Level in the Environment Associated with Climate Changes. *Appl. Sci.* **2020**, *10*, 8664. [CrossRef]
2. Cossa, D.; Tabard, A.-M. Mercury in Marine Mussels from the St. Lawrence Estuary and Gulf (Canada): A Mussel Watch Survey Revisited after 40 Years. *Appl. Sci.* **2020**, *10*, 7556. [CrossRef]
3. Brigham, M.E.; VanderMeulen, D.D.; Eagles-Smith, C.A.; Krabbenhoft, D.P.; Maki, R.P.; DeWild, J.F. Long-Term Trends in Regional Wet Mercury Deposition and Lacustrine Mercury Concentrations in Four Lakes in Voyageurs National Park. *Appl. Sci.* **2021**, *11*, 1879. [CrossRef]

4. Hernández-Fernández, J.; López-Barrera, E.A.; Mariño-Ramírez, L.; Rodríguez-Becerra, P.; Pinzón-Velasco, A. Oxidative Stress Biomarkers in Erythrocytes of Captive Pre-Juvenile Loggerhead Turtles Following Acute Exposure to Methylmercury. *Appl. Sci.* **2020**, *10*, 3602. [CrossRef]
5. Samaniego, J.; Gibaga, C.R.; Tanciongco, A.; Rastrullo, R. Total Mercury in Soils and Sediments in the Vicinity of Abandoned Mercury Mine Area in Puerto Princesa City, Philippines. *Appl. Sci.* **2020**, *10*, 4599. [CrossRef]
6. Guzmán-Uria, F.; Morales-Belpaire, I.; Achá, D.; Pouilly, M. Particulate Mercury and Particulate Organic Matter in the Itenez Basin (Bolivia). *Appl. Sci.* **2020**, *10*, 8407. [CrossRef]
7. Hellal, J.; Schäfer, J.; Vigouroux, R.; Lanceleur, L.; Laperche, V. Impact of Old and Recent Gold Mining Sites on Mercury Fluxes in Suspended Particulate Matter, Water and Sediment in French Guiana. *Appl. Sci.* **2020**, *10*, 7829. [CrossRef]
8. Guédron, S.; Achá, D.; Bouchet, S.; Point, D.; Tessier, E.; Heredia, C.; Rocha-Lupa, S.; Fernandez-Saavedra, P.; Flores, M.; Bureau, S. Accumulation of Methylmercury in the High-Altitude Lake Uru Uru (3686 m asl, Bolivia) Controlled by Sediment Efflux and Photodegradation. *Appl. Sci.* **2020**, *10*, 7936. [CrossRef]
9. Couic, E.; Alphonse, V.; Livet, A.; Giusti-Miller, S.; Bousserrhine, N. Influence of Ecological Restoration on Mercury Mobility and Microbial Activities on Former Guyanese Mining Sites. *Appl. Sci.* **2021**, *11*, 2231. [CrossRef]

applied sciences

MDPI

Article

Mercury in Hair of Mammoth and Other Prehistorical Mammals as a Proxy of Hg Level in the Environment Associated with Climate Changes

Stella Eyrikh [1,*], Gennady Boeskorov [2], Tatyana Serykh [1], Marina Shchelchkova [3] and Tatyana Papina [1]

1 Institute for Water and Environmental Problems, Siberian Branch of the Russian Academy of Science SB RAS, 656038 Barnaul, Russia; tangers62@gmail.com (T.S.); tanya.papina@mail.ru (T.P.)

2 Diamond and Precious Metals Geology Institute, Siberian Branch of the Russian Academy of Science, 677890 Yakutsk, Russia; gboeskorov@mail.ru

3 Biological Department, Institute of Natural Sciences, M.K. Ammosov's North-Eastern Federal University, 677000 Yakutsk, Russia; mar-shchelchkova@yandex.ru

* Correspondence: steyrikh@gmail.com; Tel.: +7-385-224-8007

Received: 5 October 2020; Accepted: 30 November 2020; Published: 3 December 2020

Abstract: The paper presents the first results of Hg determination in the hair of prehistorical animals (woolly mammoth, steppe bison, and woolly rhino). Hair of prehistorical mammals can be used as an archive that preserves changes of environmental pollution at the paleoscale. The aim of our study was to assess the levels of Hg exposure of ancient animals and to understand whether Hg concentration in hair could be used as a proxy indicating changes of mercury levels in the environment following global climate changes. We assessed changes of Hg exposure recorded in hairs of seven specimens of mammoth fauna mammals that inhabited the Yakutia region in the period from 45 to 10 ka yr BP. Hg concentrations in hair varied from 0.017 to 0.177 μg/g; the lowest Hg concentration were determined in older specimens (45–33 kyr yr BP). The two highest concentrations belonged sample from the Last Glacial Maximum and the Karginian interstadial (57–24 kyr BP) periods. Our hypothesis is the increase of Hg concentrations in hair reflecting environmental Hg level might be forced by high dust load in cold periods and thawing permafrost in warm climatic periods. Long-term variations of Hg level recovered from Ice Age animals' hair correlate with Hg profiles of concentration and deposition reconstructed from the Antarctica ice core.

Keywords: mercury; mammoth fauna mammals; hair; environmental changes; paleoclimate; Pleistocene; Yakutia

1. Introduction

The content of macro- and microelements in human and animal hairs is a good indicator of their accumulation in the body as a result of environmental exposure, including intake with food and water [1,2]. Hair records the levels of toxic (lead, cadmium, arsenic, etc.) and vital elements (zinc, selenium, iron, etc.), reflecting the elemental status of the whole organism. Hair analysis is used for evaluation of health state, metabolic disorders, mineral maintenance of human and animals, and also the ecological state of the territory where they live [3–5]. The level of toxic metals in the environment indicates a potential risk for the ecosystem and for human and animal health because of bioaccumulation of some metals (particularly mercury) in the body [6,7]. The World Health Organization (WHO) recommends using hair as major biological material for testing the pollution of the human body by heavy metals, since sampling, storage, and analysis of hair samples are easier than they are for other biological materials [8]. The International Atomic Energy Agency (IAEA) uses hair for the monitoring of global changes in element levels in the environment worldwide [9–11]. Ancient hairs are also "keepers of history" [12–14] which help to assess the level of environmental pollution by the degree of pollutants impact on the body during previous epochs. Hairs of prehistoric animals can be the key to understanding the relationship of environmental changes with climate. The aim of our study was to understand whether Hg concentration in the hair of prehistoric mammals could be used as a proxy indicating that changes of mercury level in the environment reflect climate changes during the Late Pleistocene–Early Holocene. Permafrost and substantial precipitation are the deciding factors in preservation of mammoth soft tissues and hair over tens of thousands of years. Hairs of the woolly mammoth are studied very actively nowadays for decoding and sequencing DNA [15–17]; revealing biologic rhythms [18]; determining the type of nutrition from the balance of stable isotopes of nitrogen, carbon, and phosphorus [19,20]; and their response to short-term (seasonal) environmental changes [21]. It is therefore surprising that studies of trace elements in the hair of mammoth fauna mammals have not been done until now. There are some studies of trace element in museum samples of animal hair [22] and bird feathers [23,24], and seal hairs from a lake sediment core spanning the past 2000 years [25]. The results of these studies are useful for assessing environmental changes and anthropogenic impacts on the environment. However, most of the studies cover the span of the last thousand years, whereas the analysis of mammoth fauna mammals' hair provides a unique opportunity to evaluate environmental changes that were happening tens of thousands of years ago, during different climatic stages. It should be noted that there is a potential problem with reliability of analytical data associated with possible contamination or loss of elements during storage and analysis of samples [26]. Therefore, the development of methodological details of sample preparation and analysis of the prehistoric animals' hair require a special attention.

2. Materials and Methods

2.1. Study Sites

All the studied fossil mammals were discovered on the territory of Yakutia (Eastern Siberia, Russia) (Figure 1).

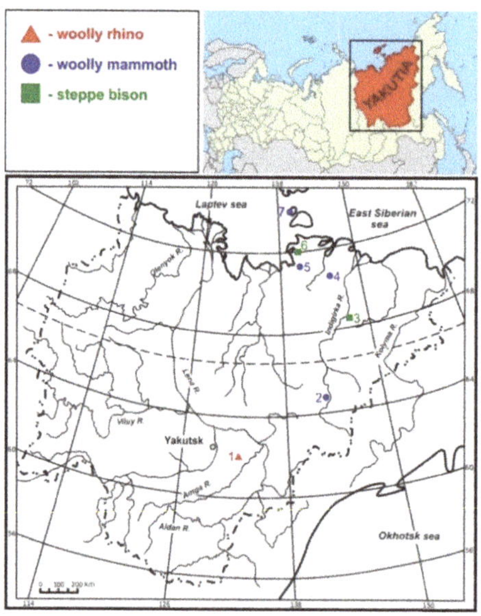

Figure 1. Study area map with location of fossil mammals discovered. Study sites were numbered from south to north.

For thousands of years, special climatic and geological conditions prevailed in the territory of Yakutia, the harshest climate in Eurasia: long cold season; low-temperatures anaerobic swamps and floodplains of rivers and lakes; gradual accumulation of precipitation on the floodplains; and permafrost growth [27,28]. These conditions allowed for the preservation of unique ancient specimens of mammals in permafrost deposits.

We studied seven specimens of Ice Age mammals; their detailed description (location and time of the discovery, estimation of the geological age by radiocarbon dating, sex, and physiological age of the animal) is presented in Table 1. The calibration of radiocarbon data was carried out by using the IntCal20 calibration curve, the version of program OxCal 4.4 (https://c14.arch.ox.ac.uk/oxcal/OxCal.html).

Table 1. Description of the studied Ice Age mammals.

Species	**Woolly Mammoth**				**Woolly Rhino**	**Steppe Bison**	
Specimen of the fossil	Berelyekh mammoth	Yukagir mammoth	Oymyakonsky baby	Malolyakhovsky mammoth	Churapcha rhino	Yukagir bison	Mylakhchin bison
Number on the map (Figure 1)	4	5	2	7	1	6	3
Site and year of recovery	Berelyekh Mammoth Cemetery, 1970	Near Yukagir settlement, Ust'-Yanskiy district, 2003	Ol'chan river, Oimyakonskiy district, 2004	The Malyi Lyakhovskiy Island, 2013	Lena-Aldan interfluve, Churapcha settlement, 1972	Chukchalakh lake bank, Ust'-Yansky district, 2011	Indigirka River, Mylakhchin locality, 1971
^{14}C yr BP	13,700 ± 80 (MAG-114)	18,510 ± 80 (GrN-28258) 18,680±100 (GrN-24288)	41,300 ± 900 (GrA-30727)	28,660 ± 160 (GrA-60021) 28,570 ± 150 (GrA-60021)	19,500 ± 120 (GIN-9594)	9295 ± 45 (GrA-53292)	29,560 ± 100 (SOAN-1007)
Cal yr BP, range, probability	16,901 – 16,321 (89.1%)	22,907 – 22,307 (95.4%)	45,735 – 42,904 (95.4%)	33,655 – 32,065 (95.4%)	23,792 – 23,151 (95.4%)	10,588 – 10,336 (95.4%)	34,398 – 33,878 (95.4%)
Gender, age	adult female	adult male, 45–50 years	baby female, 14–16 month	adult female, 55 years	adult female, 30 years	adult young male, 4.1–4.5 years	young female, 2.5 years
Climatic stage * Siberian/European	LG	LGM	KARG/ HAS	KARG/ DEN	LGM	Preboreal warming (PBO)	KARG/ DEN
Reference	[29]	[30]	[31]	[32]	[33]	[34]	[35]

* Climatic stages are explained in Section 2.2. LG, Late Glacial; LGM, Last Glacial Maximum; KARG, Karginian interstadial; HAS, Hasselo stadial; DEN, Denekamp interstadial.

2.2. Features of Discovered Species and Individuals

The woolly mammoth (*Mammuthus primigenius*) is an extinct species that lived during the Pleistocene, until its extinction in the early Holocene epoch. Woolly mammoths lived in open grassland biomes, the mammoth steppe. High-productivity grasses, herbs, and shrubs dominated there. Stomach contents clearly show that the diet of the woolly mammoth was mainly grasses and sedges [36], although tree bark and twigs also constituted a small part of their winter diet [37,38]. Yukagir mammoth (Figure 2a) is an old male who lived during the Last Glacial Maximum (LGM), the maximum Sartanian glaciations in Siberia. He died by falling into a hole. Malolyakhovsky mammoth is an old female discovered on the Maly Lyakhovsky Island (New Siberian Islands in the Laptev Sea). Radiocarbon dating of bones and hairs demonstrated good agreement, attributing the lifetime of the mammoth to the Karginian interstadial (KARG). It was a relatively warm climatic phase in Late Pleistocene in Siberia (57–24 kyr BP), which overlaps with the Denekamp interstadial warming in Europe. Berelyokh mammoth was found in one of the biggest fossil sites, the Berelyokh Mammoth Cemetery located in the basin of the Indigirka River. The entire back leg, 175 cm long, was discovered (Figure 2c), with the longest hairs reaching 120 cm. Probably it belongs to the adult female which lived in the Late Glacial (LG). Oymyakon mammoth is a baby female; she died by falling into a permafrost crack. Only the upper part of her body is preserved well. This is the oldest of the samples from the beginning of the Karginian interstadial in Siberia (Hasselo stadial in European classification). Stocky limbs and thick wool of the woolly rhino (*Coelodonta antiquitatis*) are well suited to the cold and arid steppe–tundra environment prevalent during the Pleistocene glaciations. "Churapcha rhino" (Figure 2d) is an adult female who lived during the Last Glacial Maximum (LGM); she died by falling into a coastal swamp shortly before it froze.

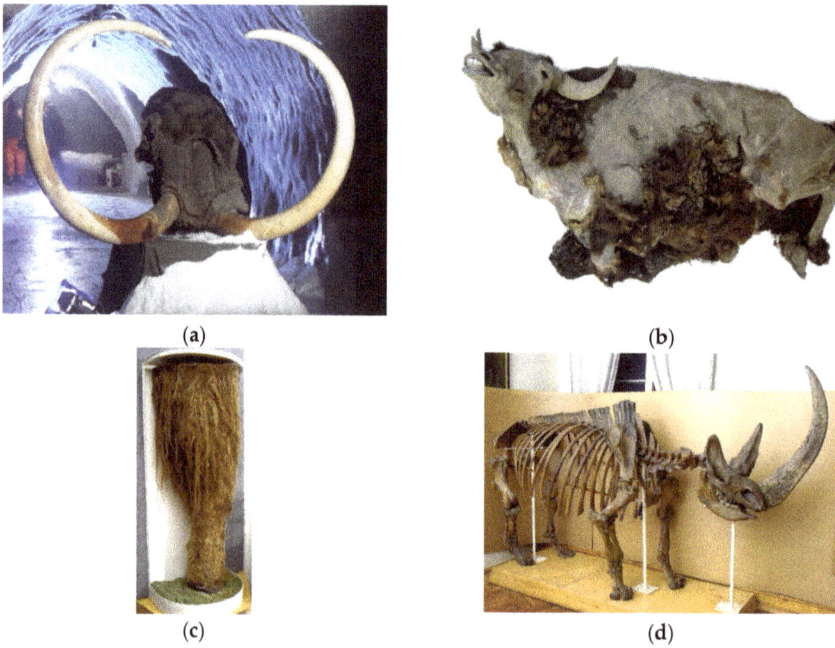

Figure 2. Photographs of the studied fossils of mammoth fauna mammals: (**a**) Yukagir mammoth head, (**b**) Yukagir bison, (**c**) posterior leg of Berelyokh mammoth, and (**d**) Churapcha rhino skeleton.

The steppe bison (*Bison priscus*), the mammoth, and the woolly rhino were the last largest herbivores that survived in Eurasia by the turn of the Pleistocene and Holocene. Steppe bison ate sedges, cereals, and plants from the forbs group. Fossil bison mummies are extremely rare. Only four well-preserved corpses are known, with two of them having been discovered in Yakutia: Yukagir and Malykhchin bison. The Yukagir bison (Figure 2b) is a young male, a complete frozen mummy that was the youngest of the studied fossil belonging to early Holocene, confirming that the bison survived an abrupt climate change at the Pleistocene–Holocene Boundary. The Malykhchin bison is a young female found on the right bank of the Indigirka River, where she died by getting stuck in coastal mud in the summer. She ate forest herbs, branches, and foliage. The presence of mosses in her stomach indicates the existence of wetland biotopes during the life of bison in the warm period of the Karginian interstadial.

Isotopic biogeochemistry helps to reveal the ecological structure of the mammoth steppe fauna. Isotopic differences reflect different dietary choices by herbivores. Woolly rhinoceros and bison grazed fresh grass, and mammoths consumed dry grass. Despite some differences in nitrogen and carbon isotopes, woolly mammoths and woolly rhinoceros are considered globally similar in diet (grass) and physiology (monogastric) [19]. Thus, it can be assumed that the differences in the concentration of mercury between the hairs of different animals reflect the changes in concentration of Hg in the environment.

Hair samples of woolly mammoth, woolly rhino, and steppe bison were obtained from the Geological Museum of Diamond and Precious Metals Geology Institute (Yakutsk). Hair records the cumulative exposure to mercury in the short- to medium-term, depending on the length of the hair sample. Whereas human hair growth rate is about 1 cm per month, and the concentration of metal in hair can show the level of mercury exposure that has occurred over many years, the animal's hair is replaced every 1–1.5 year (it holds for most of animals, both ancient and modern). Therefore, the full length of hair represents a continuous record of the elements intake over this period. Mammoth hair grows approximately 31 cm/year; the longest hair ever found covers 39 month of a mammoth's life [18,21]. Thus, the hair of mammoths and other mammoth fauna mammals reflects the environmental situation in the last years of their lives.

2.3. Sample Preparation

The determination of mercury in the hair of prehistoric animals and museum exhibits is associated with a number of difficulties: obtaining a representative sample, choosing an appropriate method given the small amount of sample, and reliable analytical determination of Hg concentration in it. Loss of volatile Hg and sample contamination are possible during long-term storage and transportation of the sample. Samples of studied fossil animals' hair were stored in museum in glass cases or wooden boxes in conditions excluding their mercury contamination during storage and therefore, they are suitable for analysis.

The amount of prehistoric animals' hair is very limited, and a single procedure of sample preparation must be developed not only for Hg, but also for a wide range of other trace elements. Here we aimed to determine the total concentrations of mercury in the hair of prehistoric animals both endogenous and exogenous in origin reflecting the intake from food and water, as well as from the air.

Methylmercury easily incorporated into hairs as it grows and its concentration in the hair is proportional to the blood concentration. The high affinity of hair for metals is mainly due to the presence of cysteine or sulfhydryl (SH) groups [39]. Elemental mercury may also bind to the hydrophobic core of the melanin polymer in the hair structure [40]. The IAEA recommends hair washing procedure using acetone and deionized water [10]; it is not suitable for Hg because the fat and keratin structures of the hair are destroyed by acetone, which leads to the loss of endogenous mercury. Washing the hair with HCl solution can leach methyl mercury from hair samples [41]. We used a chemically inert detergent ("SYNERGETIC Baby", fragrance and color free) which removes only surface grease and dust from hair samples without disturbing their structure. All reagents were tested for Hg content and purified if necessary. Nitric and hydrochloric acids were purified using a Savillex DST-1000 distillation system (Savillex, Eden Prairie, MN, USA). Ultrapure water (MQ-water) was obtained using a Simplicity UV water purification system (Millipore SAS, Molsheim, France). All stages of sample preparation were carried out in a "clean room" equipped with outdoor air handlers that use progressively finer filters including high-efficiency particulate air (HEPA) filter and charcoal mercury filter, which remove particulate and elemental mercury from the incoming air.

The washing procedure for hair comprises the following steps:

(1) Washing the samples in a solution of chemically inert detergent (30 min);
(2) Three-fold soaking in MQ-water (during the day);
(3) Air-drying in a clean room;
(4) Cutting hair into small pieces of less than 0.3 cm (scissors pretreated with detergent and ultrapure water);
(5) Careful homogenization of the obtained samples by mixing to provide a representative sub-sampling;
(6) Storage in a double-sealed zipper-locked plastic bag, in a clean room.

Microwave system MARS-5 (Thermo Fisher Scientific, Waltham, MA, USA) was used for digestion of hair samples by the program previously optimized for Hg analysis in biological objects (Table 2) [42].

We tested acid and acid–peroxide digestion and demonstrated applicability of acid–peroxide digestion (2 mL HNO_3 + 1 mL H_2O_2) for Hg analysis. The latter method was used for the analysis of samples.

Table 2. Optimized parameters of microwave digestion of biotic samples.

Stage	Power, W/%	Time, min	Pressure, psi	Temperature, °C	Retention Time, min
1	1200/100	5	20	85	5
2	1200/100	5	80	140	0
3	1200/100	5	160	180	0
4	1200/100	5	190	180	10

After microwave digestion the samples were cooled to 25 °C, the pressure was brought to <50 psi, and sample volume was adjusted to 10–12 mL with MQ water. The procedure of sample preparation (washing, cutting, and digestion) was developed and tested by using hair samples of modern yak living in the Barnaul Zoo and Certified Reference Material of human hair (CRM, Hair NSC DC 73347, China). The developed sample-preparation procedure is suitable for both for Hg and multi-element analysis of prehistorical animals' hair and blood.

2.4. Hg Analysis

Content of mercury in hair and blood samples was determined by Mercur Duo Plus Analyzer (Analytik Jena, Jena, Germany), combining atomic fluorescence with the cold vapor method and amalgamation on gold collector. Analytical characteristics of the method are presented in Table 3. The accuracy was confirmed by using Certified Reference Material of human hair (CRM, Hair DC 73347, China). Optimization of the instrumental parameters [43], using ultrapure reagents and clean conditions, allowed us to achieve a method detection limit of up to 0.4 ng/L for liquid samples and 0.003 μg/g for hair samples (0.03 g, dry weight). Split sampling and analyzing the same samples at different times and by different operators were used for assessing precision, recovery, and reproducibility. Good spike recovery values were demonstrated for samples of yak hair and CRM. The confidence interval for low-concentration samples did not exceed 17%.

Table 3. Analytical characteristics of method.

Element	Accuracy, μg/g		Spike Recovery, %		Precision (%)		Method Detection Limit, μg/g
	Measured Value	Certified Value [1]	CRM	Yak	Within Run	Between Runs	
Hg	0.36 ± 0.04	0.36 ± 0.08	94.7	100.4	3%	5%	0.003

[1] Certified value is for human hair, Certified Reference Material (CRM) NSC DC 73347, China.

3. Results and Discussion

3.1. Hg Concentration in Fossil Animals' Hair

Mercury concentrations determined in the prehistoric animals' hair varied from 0.017 to 0.177 μg/g; average concentrations and ranges of Hg content in different types of prehistoric and modern animals are presented in Table 4, together with reference values and intervals. Mercury coming directly from water, air, and food tends to accumulate in both plants and animals, being toxic to most life forms. WHO guidance established 2 μg/g for total Hg in human hair as the reference level for risk evaluation [44]. The US Environmental Protection Agency (EPA) sets the reference dose for human hair and wildlife toxicity at 1 μg/g [45]. All of the prehistoric animals' hairs have Hg concentrations significantly below these levels. Moreover, they do not exceed the background level of mercury in hair of non-seafood consumers (0.5 μg/g). As far as we are aware, there is no background assessment of Hg level in herbivore prehistorical animals. We can compare our results with Hg concentrations in the hair of modern herbivore animals and their reference intervals for Hg (discussed in Section 3.2).

Table 4. Hg concentrations in hair of modern and prehistoric animals and reference values.

Hg Concentration, µg/g	Prehistoric Animals			Modern Animals		Reference Values
	Mammoth	Bison	Woolly Rhinoceros	Yak	Cattle	
number of samples	4	2	1	1	145	-
hair average	0.031	0.11	0.092	0.0056	0.0066 [46]	1 [45] <0.5 [47]
hair range	0.017–0.049	0.035–0.177	-	-	0.002–0.011 [48]	0.02–0.042 [48]
blood *	0.69 ± 0.07	-	-	-	<0.438 (ND-15.4) [49]	<2 [45]

* µg/L; ND—not detected.

3.1.1. Woolly Mammoth Hair and Hemolyzed Blood

Hairs of four fossil mammoths were studied; the Hg concentrations are shown in the Figure 3a. Higher concentrations were in hair samples of Berelyekh and Yukagir mammoths from the Last Glacial and Preboreal warming, respectively, whereas the lowest Hg concentrations were in older specimens, Oymyakonsky and Malolyakhovsky mammoths (45–33 ka yr BP).

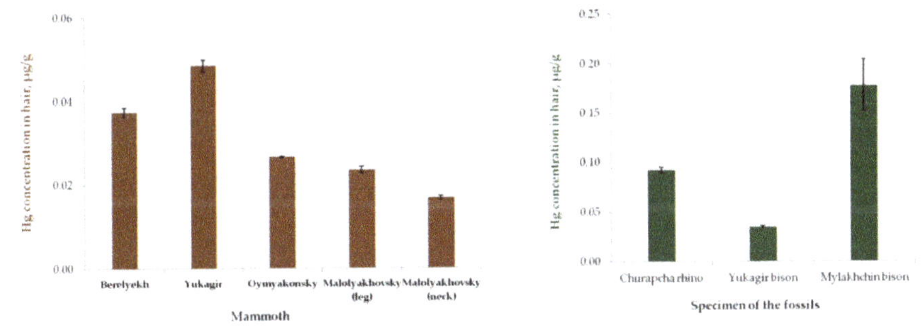

(a) (b)

Figure 3. Mercury concentration in hair of different fossil specimens: (**a**) mammoth; (**b**) bison and rhino.

There was rare opportunity to sample enough freshly thawed hair material from different parts of the body of the same animal (Malolyakhovsky mammoth). Samples of hair taken from the neck and back leg demonstrated slightly different concentrations: 0.017 and 0.024 µg/g, respectively. The difference between Hg concentrations sampled from different body areas for modern animals varied insignificantly for yak from 0.004 to 0.007 µg/g (this study) and for beef cattle from 0.062 to 0.070 µg/g [46]. The comparison of guard and down yak hair demonstrated a negligible difference in Hg concentrations: 0.0054 and 0.0058 µg/g, respectively. Cattle hair also yielded similar Hg concentrations: 0.010 and 0.008 µg/g for guard and down hairs. It was shown that the elemental composition and concentrations of most elements of beef cattle hair on different body areas does not differ notably [46]. Of course, in the ideal case, hair sampled from the same body areas should be used for the comparison of the animals. However, unfortunately, there are very few opportunities to get hair samples from the same part of body for different prehistoric animals. Therefore, comparison of prehistoric animals' hair sampled from different body surface areas is evidently acceptable due to minor differences in the concentrations as confirmed by our results (no more than 30%).

There was rare chance to determine Hg concentration in hemolyzed blood of Malolyakhovsky mammoth. Mercury measurement in the whole blood provides information about recent exposure (~1–2 months) to both organic and inorganic mercury through ingestion of food and drinking water and inhalation of elemental mercury vapor in ambient air. The level of mercury in blood indicates recent exposure, but it does not reflect historical exposure or variations in exposure. Here we determined Hg concentration in mammoth hemolyzed blood 0.69 ± 0.07 µg/L. Hg levels in mammoth blood are below background levels of mercury in the blood of people who do not consume fish

(<2 μg/L) [45]. It is comparable with relatively low levels of total mercury in the blood of modern animals: The median value of Hg in blood of Galician cows from NW Spain was <0.438 μg/L, with the range of <0.438 to 15.4 μg/L [49]. This is similar to Hg concentrations in blood of surveyed dogs from Alaska (0.16–12.38 ng/g) [50].

Hair to blood Hg ratio in Malolyakhovsky mammoth was 36, which is closer to the hair to blood ratio in fish-fed dogs (59 ± 7.6), harbor seals (22–40), and polar bears (100) [51], than to the WHO value for humans (250) used for risk assessments to predict blood Hg from hair concentrations [8]. The differences in ratios may be due to the differences in relative surface area and hair density for different animal species and humans. Unfortunately, we have a single blood sample for mammoth, so we can only assume that this ratio level is characteristic for mammoths and other Ice Age animals.

3.1.2. Steppe Bison Hair

Hg concentrations in the hair of two steppe bison differ by about five times: 0.035 ± 0.001 and 0.177 ± 0.026 μg/g (Figure 3b). The concentration of Hg in Yukagir bison hair is at a comparable level to the Berelekh mammoth hair, whereas Hg concentration in hair of Mylakhchin bison is the highest among all studied Ice Age animals. We hypothesize that there was an increase in mercury concentrations in the environment during this period. Mylakhchin bison lived in Karginian interstadial, in conditions of climate warming. The latter might be responsible for enhanced release of Hg due to thawing permafrost. Reconstruction based on palynological data revealed that, during the Karginian interstadial, there were stages with warmer and milder-than-today climate conditions, and the amplitude of climate fluctuations was different for different regions of Siberia [52]. Modern bison are very similar to the prehistoric ones in terms of nutrition, wool structure, etc. It has been observed that bison can find food under deep snow layers (>50 cm) [53]. Unfortunately, we could not find mercury concentrations in the hair of modern bison, although very low levels of hepatic Hg in the liver of captive and free-ranging European Bison from two different sites (0.003 μg/g) indicate a low mercury load [54]. The levels of such vital trace elements, such as iron, titanium, and vanadium, in the hair of a modern European bison are much lower than in hair of both prehistoric fossils [55].

3.1.3. Woolly Rhino Hair

There was only one sample of woolly rhino hair, and it had a high Hg concentration of 0.092 ± 0.003 μg/g (second highest of all studied samples). This is supposedly related to Hg variability based on climatic stage and is discussed in detail in Section 3.4.

3.2. *Comparison with Modern Animals' Hair and Reference Intervals*

Unfortunately, we cannot compare Hg levels in modern animals to historical levels in the same animal species, because there are no modern animals identical to the mammoth mammals. Elephants are closest to mammoths genetically, but they have a different body and habitat. We compared the Hg concentration in hair of mammoth and yak, as they have similar characteristics (nutrition, long hair, etc.). A yak (*Bos mutus*) from the Barnaul Zoo has a low Hg level (0.006 ± 0.001 μg/g) (Table 4). There are no data about Hg concentration in the hair of yak from other regions of the world, but concentrations of other elements were found to be comparable between Altai and Asian yak, indicating that the differences in their exposure to metals are insignificant for the vast territories of their habitats [56,57]. Hg concentrations in all mammoth mammals' hair samples were significantly higher than Hg concentration in unpolluted hair samples of modern animals such as cattle (0.0066 ± 0.0002 μg/g) (Table 4). Methods for determining reference ranges in hair by using results from a large human population are described in detail elsewhere [11]. The reference intervals and 90% confidence intervals for the lower and upper limits were calculated for hair trace-element content in cattle (*Bos taurus*) per the recommendations of the American Society for Veterinary Clinical Pathology Quality Assurance and Laboratory Standard Guidelines [48,58]. Concentrations of Hg in mammoth fauna mammals' hair mostly lie within the optimal reference range for cattle (Table 4), excluding two highest concentrations, which apparently reflect high environmental exposure to mercury in these mammals during the last periods of their lives. For plant-eating animals, vegetation is one of the main factors characterizing the living conditions [20], although Hg accumulation by animals depends both on their diet and habitat.

3.3. Hg Levels in Arctic Animals and Humans (Historical and Modern)

In historical samples of hair of human mummies of the Aleutian Islands (Alaska) dating 1450 AD, mean total mercury concentration (5.8 ± 0.9 μg/g) is comparable to the levels observed in hair of modern residents of the northern polar territories (Alaska, Canada, Faroe Islands) [59]. That confirms the main contribution of the traditional nutrition based on fish and meat of marine mammals to the accumulation of mercury for residents of these territories. The Egyptian, Chilean, and Peruvian mummies had mercury exposures below the US EPA reference level of 1 μg/g and were considerably lower than that of northern pre-industrial populations [60]. Hg concentrations in hair samples of historical (10.42 ± 1.31 μg/g) and modern (10.42 ± 2.45 μg/g) arctic foxes were similar and strongly correlated with ecotype and available food source [22]. Unlike humans and foxes, Hg concentrations in the hair of Greenland polar bears showed a significant increase from 0.52 to 4.9 μg/g (from 1300 to 2000 years) [61]. Comparison between Hg levels in the hair of the ancient dogs of the Seward Peninsula (0.657 ± 0.273 ng/g [62]) and the modern Alaska fish-fed dogs (0.54 ± 0.11 μg/g, [51]) did not show significant difference. Thus, when environmental exposure (atmosphere and water) to mercury is low, the increased levels of mercury in the bodies of ancient humans and animals are primarily associated with their diet. Mammals of the mammoth fauna have low Hg levels, since they are herbivores that get mercury from plants and accumulate it in their bodies (and hair). Biomagnification along the food chain (as seen in aquatic ecosystems and fish-eating animals and humans) is not observed.

3.4. Comparison with Other Paleoarchive Data

Environmental archives such as lake and marine sediments, peat bogs, glacial ice, and tree rings are widely used to reconstruct Hg accumulation at the local, regional, and global scale. All archives have their advantages and disadvantages, but none of them is a definite record of past mercury levels, because of the complexity of the mercury cycle's being influenced by various processes in each archive [63,64]. Most archives record the past several hundred to several thousand years (ice cores, peat bogs, lake and marine sediment cores, and tree rings), whereas long-term paleorecords recording up to a hundred thousand years are scarce (ice cores of Antarctica and Greenland, sediments cores and speleothems). Ice Age animals in this study lived 45 up to 10.5 kyr BP. Animals were exposed to mercury from the diet and the environment. The natural sources of Hg emissions were volcanoes, air–sea and soil–vegetation–air exchange, biomass burning (wildfires), and the revolatilization of deposited Hg from the soils (including release associated with permafrost thawing due to climate change).

The highest Hg concentration recorded in this study dates to 33.930 cal kyr BP. The other peak of Hg concentration at 23.292 cal kyr BP falls into the LGM period. An increase of mercury concentrations in hair coincides with variations of Hg concentrations and depositions recorded in the Antarctica Dome C ice core (Figure 4). Due to constant snow accumulation, Hg concentrations and fluxes change synchronically. Total Hg and Hg^{2+} concentrations are also characterized by similar trends, except during the initial period from 15 to 2 kyr BP. The highest peak occurred during the Karginian interstadial of the Late Pleistocene, the period of maximum insolation in 200 ka years [65]. The presence of mosses in Malykhchinsky bison's food masses indirectly indicated a significant wetland area in this period because of significant climate warming. Hg increase in environment caused by rapid release of mercury during thawing periods was recorded in other paleoarchives, such as sediments in Limnopolar Lake (South Shetland Islands), where extraordinary high Hg enrichment was observed [65]. Research based on about 600 samples from soil permafrost cores (Alaska) discovered that the active layer is the largest Hg pool on the planet. The Northern Hemisphere permafrost region contains 1656 ± 962 Gg Hg, of which 793 ± 461 Gg is frozen in permafrost. The active layer and permafrost contain nearly twice as much Hg as all other soils, the ocean, and the atmosphere combined [66].

This allows us to assume that, in the past warm climatic periods, thawing permafrost caused significant mercury to be released into the environment, from the active layer of permafrost.

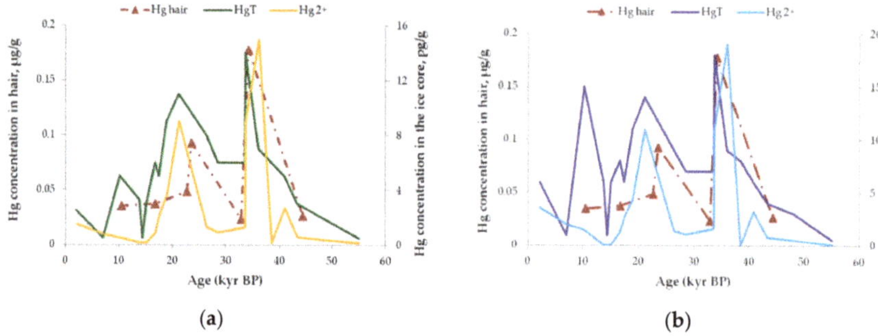

Figure 4. Concentrations (**a**) and fluxes (**b**) of total mercury (HgT) and inorganic mercury (Hg^{2+}) in the EPICA Dome C ice core [67] and total Hg concentrations in mammoth fauna mammals' hair. Concentrations and fluxes of Hg below the Method Detection Limit (MDL) are presented in the graph as 1/2 of MDL.

The exogenous Hg in hair of ancient animals is mainly due to its sorption on the hair surface, from the atmosphere. The endogenous Hg in hair also can be due to Hg^0 influx. Elemental mercury (up to 80% of inhaled Hg^0 vapors) is absorbed in the lungs, quickly diffused into the blood, and distributed to all organs of the body; it also accumulates in growing hair.

The second peak corresponds to the Last Glacial Maximum, where the Antarctic record also shows a drastic increase in Hg concentrations during the LGM. It was found for mercury in the Antarctic [67] that the oxidation of gaseous mercury by sea-salt-derived halogens occurred in the cold atmosphere. The oxidized mercury compounds were then transferred to the abundant mineral dust particles and deposited. A significant correlation between dust concentrations and changes of temperature during glacial periods was confirmed by comparing dust and stable isotope, up to 90% of the dust variability can be explained by the temperature variations. The deposition of dust in Antarctica during glacial periods is about 20 times higher than during interglacials [68]. The cooling marked in three independently dated North Atlantic marine sediment cores is synchronous with the sharp increase in dust flux recorded in the Greenland ice cores, an increase in dust transport from Asia to Greenland observed during few Greenland stadials [69,70]. Deposition of mercury with dust on the surface of the land and its accumulation by the plants and snow might be the reason of increased dietary Hg exposure of herbivores. A high content of loamy particles in fecal samples of mammoths indicates an occasional or deliberate lithophagy [71]. Moreover, when thick ice completely covers the water, animals eat snow. Thus, changes of mercury concentrations in the hair of prehistoric animals are in good agreement with global changes of mercury concentrations recorded in other paleoarchives of Northern Hemisphere. It should be noted that the question remains open: Is there is a real difference in the deposition, distribution, and conservation of mercury in the Northern and Southern Hemispheres, or are Hg changes global? All archives preserve Hg differently and present changes in global Hg cycle at various spatial and temporal scales [64].

4. Conclusions

Mercury content in the hair of mammoths and other prehistoric animals allows us to estimate changing mercury levels between 40,000 to 10,000 years ago. Since the amount of ancient hair is very limited, we suggest comparing the hair of different animal species which are similar in diet and habits, as well as hair samples taken from different parts of the animals' body. All prehistoric animals have a low Hg level in their hair. This level is below concentrations associated with toxicity in wildlife and do not exceed background levels of mercury in hair of non-seafood consumers (0.5 μg/g). Most of the Hg concentrations in the hair of prehistoric animals were within the reference range for modern cattle.

There are many advantages to using ancient hair as an indicator of environmental pollution, and now we present a new application of hair as an indicator of climatic changes. We hypothesize that Hg concentrations in hair reflect the variation in Hg level in the environment changing with climate changes, and can be used as a proxy for climate change assessment. The increase of Hg concentration in hair during the coldest climatic stages (such as LGM) coincides with the increase in Hg deposition on the Earth's surface, associated with the highest atmospheric dust loads. Moreover, mercury can be released to the atmosphere because of permafrost thawing during interstadial warming; the highest Hg concentration coincides with the Karginian interstadial of the Late Pleistocene, the period of maximum insolation and warming. Climate changes in warm and cold climatic stages were oscillatory; relatively warm periods alternated with cooler periods during each glacial and interglacial. For example, Karginian interstadial consisted of five periods (three warming and two cooling), in which features of the distribution and boundaries of permafrost are still under study.

Mammoth fauna mammals' hair, together with other natural archives, will be useful in assessing the response of Hg cycle to climate change. More paleo data are necessary to confirm our first finding, and to clarify whether these changes will differ for the Northern and Southern Hemispheres' archives, so we are planning further studies of mammoth fauna mammals on a wide spatiotemporal scale.

Author Contributions: This work was carried out in the framework of cooperation between Institute for Water and Environmental Problems from Barnaul, Diamond and Precious Metals Geology Institute, and Ammosov's North-Eastern Federal University from Yakutsk. Conceptualization and investigation, S.E. and G.B.; mammoth samples and morphological information, G.B.; methodology, S.E.; investigation and analysis, T.S., S.E., and M.S.; supervision T.P.; resources and funding acquisition, G.B. and T.P.; writing—original draft preparation, S.E.; writing—review and editing, S.E., G.B., and T.P. All authors have read and agreed to the published version of the manuscript.

Funding: This research was performed under a State Assignment of the IWEP SB RAS (project no. AAAA-A17-117041210242-1). Studies by G.G. Boeskorov were performed under a State Assignment of the Institute of Diamond and Precious Metal Geology, Siberian Branch, Russian Academy of Sciences (project no. 0381-2019-0002); studies by M.V. Shchelchkova were performed under a State Assignment of the Ministry of Education and Science of the Russian Federation (project no. 37.7935.2017/6.7).

Acknowledgments: We are grateful to the Director of the Barnaul Zoo Sergey Pisarev for his help in obtaining yak hair samples for method testing. We expressed our deep gratitude to Eugenia Bakunova for kind help in English editing of the manuscript.

Conflicts of Interest: The authors declare no conflict of interest.

References

1. Airey, D. Mercury in human hair due to environment and diet: A review. *Environ. Health Perspect.* **1983**, *52*, 303–316. [CrossRef] [PubMed]
2. Hejna, M.; Gottardo, D.; Baldi, A.; Dell'Orto, V.; Cheli, F.; Zaninelli, M.; Rossi, L. Review: Nutritional ecology of heavy metals. *Animal* **2018**, *12*, 2156–2170. [CrossRef] [PubMed]
3. Combs, D.K.; Goodrich, R.D.; Meiske, J.C. Mineral concentrations in hair as indicators of mineral status—A review. *J. Anim. Sci.* **1982**, *54*, 391–398. [CrossRef] [PubMed]
4. Katz, S.A.; Chatt, A. *Hair Analysis: Applications in the Biomedical and Environmental Sciences*; VCH Publishers: New York, NY, USA, 1988.
5. Jung, R.-S.; Yang, S.-R.; Han, J.K.; Kang, G.H.; Lee, G.H. Determination of Lead, Cadmium, and Chromium in Hair Optimized by Simplex Method Using Electrothermal Vaporization-Inductively Coupled Plasma Mass Spectrometry. *Anal. Sci. Suppl.* **2002**, *17icas*, i999–i1002. [CrossRef]

6. Pavlogeorgatos, G.; Kikilias, V. The importance of mercury determination and speciation to the health of the general population. *Glob. Nest Int. J.* **2002**, *4*, 107–125.
7. Okpala, C.O.R.; Sardo, G.; Vitale, S.; Bono, G.; Arukwe, A. Hazardous properties and toxicological update of mercury: From fish food to human health safety perspective. *Crit. Rev. Food Sci. Nutr.* **2018**, *58*, 1986–2001. [CrossRef]
8. *Human Biomonitoring: Facts and Figures*; WHO Regional Office for Europe: Copenhagen, Denmark, 2015.
9. *Mercury Contamination in Man and His Environment*; International Atomic Energy Agency: Vienna, Austria, 1972.
10. International Atomic Energy Agency, Agency's Laboratories, Analytical Quality Control Services. *Activation Analysis of Hair as an Indicator of Contamination of Man by Environmental Trace Element Pollutants*; Ryabukhin, Y.S., Ed.; International Atomic Energy Agency (IAEA): Seibersdorf, Austria, 1976; p. 135.
11. Druyan, M.E.; Bass, D.; Puchyr, R.; Urek, K.; Quig, D.; Harmon, E.; Marquardt, W. Determination of reference ranges for elements in human scalp hair. *Biol. Trace Elem. Res.* **1998**, *62*, 183–197. [CrossRef]
12. Toribara, T.Y.; Muhs, A.G. Hair: A Keeper of History. *Arct. Anthropol.* **1984**, *21*, 99–108.
13. Egeland, G.M.; Ponce, R.; Knecht, R.; Bloom, N.S.; Fair, J.; Middaugh, J.P. Trace metals in ancient hair from the Karluk Archaeological Site, Kodiak, Alaska. *Int. J. Circumpolar Health* **1999**, *58*, 52–56.
14. Thompson, A.H.; Wilson, A.S.; Ehleringer, J.R. Hair as a geochemical recorder: Ancient to modern. In *Treatise on Geochemistry*, 2nd ed.; Turekian, K.K., Holland, H.D., Eds.; 14—Archaeology & Anthropology; Elsevier: Amsterdam, The Netherlands, 2014; pp. 371–393. [CrossRef]
15. Gilbert, M.T.P.; Tomsho, L.P.; Rendulic, S.; Packard, M.; Drautz, D.I.; Sher, A.; Tikhonov, A.; Dalen, L.; Kuznetsova, T.; Kosintsev, P.; et al. Whole-genome shotgun sequencing of mitochondria from ancient hair shafts. *Science* **2007**, *317*, 1927–1930. [CrossRef]
16. Wong, K. Decoding the Mammoth Scientists sequence half the woolly mammoth's genome. *Sci. Am.* **2009**, *300*, 26–27.
17. Roca, A.L.; Ishida, Y.; Nikolaidis, N.; Kolokotronis, S.-O.; Fratpietro, S.; Stewardson, K.; Hensley, S.; Tisdale, M.; Boeskorov, G.; Greenwood, A.D. Genetic variation at hair length candidate genes in elephants and the extinct woolly mammoth. *BMC Evol. Biol.* **2009**, *9*, 232. [CrossRef] [PubMed]
18. Spilde, M.; Lanzirotti, A.; Qualls, C.; Phillips, G.; Ali, A.M.; Agenbroad, L.; Appenzeller, O. Biologic Rhythms Derived from Siberian Mammoths2019 Hairs. *PLoS ONE* **2011**, *6*, e21705. [CrossRef] [PubMed]
19. Bocherens, H. Isotopic tracking of large carnivore palaeoecology in the mammoth steppe. *Quat. Sci. Rev.* **2015**, *117*, 42–71. [CrossRef]
20. Ukraintseva, V. *Vegetation and Climate of Siberia in the Mammoth Epoch*; East Siberian Branch of the International Institute of Forest: Krasnoyarsk, Russia, 2002. (In Russian)
21. Iacumin, P.; Davanzo, S.; Nikolaev, V. Short-term climatic changes recorded by mammoth hair in the Arctic environment. *Palaeogeogr. Palaeoclimatol. Palaeoecol.* **2005**, *218*, 317–324. [CrossRef]
22. Bocharova, N.; Treu, G.; Czirjak, G.A.; Krone, O.; Stefanski, V.; Wibbelt, G.; Unnsteinsdottir, E.R.; Hersteinsson, P.; Schares, G.; Doronina, L.; et al. Correlates between Feeding Ecology and Mercury Levels in Historical and Modern Arctic Foxes (*Vulpes lagopus*). *PLoS ONE* **2013**, *8*, e60879. [CrossRef]
23. Vo, A.-T.E.; Bank, M.S.; Shine, J.P.; Edwards, S.V. Temporal increase in organic mercury in an endangered pelagic seabird assessed by century-old museum specimens. *Proc. Natl. Acad. Sci. USA* **2011**, *108*, 7466–7471. [CrossRef]
24. Strekopytov, S.; Brownscombe, W.; Lapinee, C.; Sykes, D.; Spratt, J.; Jeffries, T.E.; Jones, C.G. Arsenic and mercury in bird feathers: Identification and quantification of inorganic pesticide residues in natural history collections using multiple analytical and imaging techniques. *Microchem. J.* **2017**, *130*, 301–309. [CrossRef]
25. Sun, L.; Yin, X.; Liu, X.; Zhu, R.; Xie, Z.; Wang, Y. A 2000-year record of mercury and ancient civilizations in seal hairs from King George Island, West Antarctica. *Sci. Total Environ.* **2006**, *368*, 236–247. [CrossRef]
26. Pozebon, D.; Scheffler, G.L.; Dressler, V.L. Elemental hair analysis: A review of procedures and applications. *Anal. Chim. Acta* **2017**, *992*, 1–23. [CrossRef]
27. Popov, A.I. A mammoth from Taymyr and problem of the mammoth fauna remnants in the Siberian Quaternary deposits. In *The Ice Age in the European Section of the USSR and in Siberia*; Markov, K.K., Popov, A.I., Eds.; MGU: Moscow, Russia, 1956; pp. 259–275.
28. Gavrilova, M.K. *Climates of the Cold Regions of the Earth*; Siberian Branch of RAS Publishing House: Yakutsk, Russia, 1998.

29. Vereshchagin, N.K. Berelekh "cemetery" of mammoths. *Proc. Zool. Inst. Acad. Sci. USSR* **1977**, *72*, 5–50.
30. Boeskorov, G.G.; Tikhonov, A.N.; Suzuki, N. *The Yukagir Mammoth*; Saint-Petersburg University Publishing House: Saint-Petersburg, Russia, 2007; p. 252.
31. Boeskorov, G.G.; Tikhonov, A.N.; Lazarev, P.A. A new find of a mammoth calf. *Dokl. Biol. Sci.* **2007**, *417*, 480–483. [CrossRef] [PubMed]
32. Grigoriev, S.E.; Fisher, D.C.; Obadă, T.; Shirley, E.A.; Rountrey, A.N.; Savvinov, G.N.; Garmaeva, D.K.; Novgorodov, G.P.; Cheprasov, M.Y.; Vasilev, S.E.; et al. A woolly mammoth (*Mammuthus primigenius*) carcass from Maly Lyakhovsky Island (New Siberian Islands, Russian Federation). *Quat. Int.* **2017**, *445*, 89–103. [CrossRef]
33. Lazarev, P.A.; Boeskorov, G.G.; Tomskaya, A.I.; Garutt, N.V.; Vasilyev, E.M.; Kasparov, A.K. *Mammals of the Anthropogen of Yakutia*; Publishing House YAC SB RAS: Yakutsk, Russia, 1998.
34. Boeskorov, G.G.; Potapova, O.R.; Protopopov, A.V.; Plotnikov, V.V.; Agenbroad, L.D.; Kirikov, K.S.; Pavlov, I.S.; Shchelchkova, M.V.; Belolyubskii, I.N.; Tomshin, M.D.; et al. The Yukagir Bison: The exterior morphology of a complete frozen mummy of the extinct steppe bison, Bison priscus from the early Holocene of northern Yakutia, Russia. *Quat. Int.* **2016**, *406*, 94–110. [CrossRef]
35. Lazarev, P.A. *Large Mammals of the Antropogene of Yakutia*; Nauka: Novosibirsk, Russia, 2008.
36. Ukraintseva, V.V. *Vegetation Cover and Environment of the "Mammoth Epoch" in Siberia*; Mammoth Site of Hot Springs: Hot Springs, SD, USA, 1993.
37. Vereshchagin, N.K.; Baryshnikov, G.F. 16—Paleoecology of the mammoth fauna in the Eurasian Arctic. In *Paleoecology of Beringia*; Hopkins, D.M., Matthews, J.V., Schweger, C.E., Young, S.B., Eds.; Academic Press: Cambridge, MA, USA, 1982; pp. 267–279.
38. Olivier, R.C.D. 18—Ecology and behavior of living elephants: Bases for assumptions concerning the extinct woolly mammoths. In *Paleoecology of Beringia*; Hopkins, D.M., Matthews, J.V., Schweger, C.E., Young, S.B., Eds.; Academic Press: Cambridge, MA, USA, 1982; pp. 291–305.
39. Morton, J.; Carolan, V.A.; Gardiner, P.H.E. Removal of exogenously bound elements from human hair by various washing procedures and determination by inductively coupled plasma mass spectrometry. *Anal. Chim. Acta* **2002**, *455*, 23–34. [CrossRef]
40. Kronstrand, R.; Förstberg-Peterson, S.; Kågedal, B.; Ahlner, J.; Larson, G. Codeine concentration in hair after oral administration is dependent on melanin content. *Clin. Chem.* **1999**, *45*, 1485–1494. [CrossRef]
41. Kratzer, K.; Beneš, P.; Spěváčková, V.; Kolihová, D.; Žilková, J. Determination of chemical forms of mercury in human hair by acid leaching and atomic absorption spectrometry. *J. Anal. At. Spectrom.* **1994**, *9*, 303–306. [CrossRef]
42. Valereva, E.V.; Ilyina, E.G.; Eyrikh, S.S. Methodical features of sample preparation and determination of mercury (II) in biotic objects. In *Proceedings of Young Scientists of Altai State University*; ASU: Barnaul, Russia, 2015; pp. 109–111.
43. Eyrikh, S.; Schwikowski, M. Experience of using the atomic fluorescence analyzer "Mercur" for determination ultra-low concentrations of mercury in ice and snow samples. In *Mercury. Problems of Geochemistry, Ecology, Analytics*; Volosov, A.G., Ed.; IMGRE: Moscow, Russia, 2005; pp. 111–114.
44. *WHO/UNEP DTIE Chemical Branch Guidance for Identifying Populations at Risk from Mercury Exposure*; WHO: Geneva, Switzerland, 2008.
45. Council, N.R. *Toxicological Effects of Methylmercury*; The National Academies Press: Washington, DC, USA, 2000; p. 364. [CrossRef]
46. Miroshnikov, S.; Kharlamov, A.; Zavyalov, O.; Frolov, A.; Bolodurina, I.; Arapova, O.; Duskaev, G. Method of Sampling Beef Cattle Hair for Assessment of Elemental Profile. *Pak. J. Nutr.* **2015**, *14*, 632–636. [CrossRef]
47. AMAP/UN Environment 2019. *Technical Background Report for the Global Mercury Assessment 2018*; Arctic Monitoring and Assessment Programme: Oslo, Norway; UN Environment Programme, Chemicals and Health Branch: Geneva, Switzerland, 2019; p. 426.
48. Miroshnikov, S.A.; Zavyalov, O.A.; Frolov, A.N.; Bolodurina, I.P.; Kalashnikov, V.V.; Grabeklis, A.R.; Tinkov, A.A.; Skalny, A.V. The Reference Intervals of Hair Trace Element Content in Hereford Cows and Heifers (Bos taurus). *Biol. Trace Elem. Res.* **2017**, *180*, 56–62. [CrossRef]
49. López Alonso, M.; Benedito, J.L.; Miranda, M.; Castillo, C.; Hernández, J.; Shore, R.F. Mercury concentrations in cattle from NW Spain. *Sci. Total Environ.* **2003**, *302*, 93–100. [CrossRef]

50. Sousa, A.C.; Teixeira, I.S.; Marques, B.; Vilhena, H.; Vieira, L.; Soares, A.M.; Nogueira, A.J.; Lillebø, A.I. Mercury, pets' and hair: Baseline survey of a priority environmental pollutant using a noninvasive matrix in man's best friend. *Ecotoxicology* **2013**, *22*, 1435–1442. [CrossRef] [PubMed]
51. Lieske, C.L.; Moses, S.K.; Castellini, J.M.; Klejka, J.; Hueffer, K.; O'Hara, T.M. Toxicokinetics of mercury in blood compartments and hair of fish-fed sled dogs. *Acta Vet. Scand.* **2011**, *53*, 66. [CrossRef] [PubMed]
52. Laukhin, S.A.; Pushkar, V.S.; Cherepanova, M.V. Modern condition of environmental reconstructions on Siberian North during Karginsky time (Late Pleistocene). *Bull. Mosc. Soc. Nat. Geol. Ser.* **2012**, *87*, 37–48.
53. Vlasakker, J.V.D. *Bison Rewilding Plan 2014–2024—Rewilding Europe's Contribution to the Comeback of the European Bison*; Rewilding Europe: Nijmegen, The Netherlands, 2014.
54. Durkalec, M.; Nawrocka, A.; Krzysiak, M.; Larska, M.; Kmiecik, M.; Posyniak, A. Trace elements in the liver of captive and free-ranging European bison (*Bison bonasus* L.). *Chemosphere* **2018**, *193*, 454–463. [CrossRef]
55. Kośla, T.; Skibniewska, E.; Skibniewski, M. The state of bioelements in the hair of free-ranging European bisons from Bialowieza Primeval Forest. *Pol. J. Vet. Sci.* **2011**, *14*, 81–86. [CrossRef]
56. Chatterjee, A.; Raquib, M.; Sheikh, I.U.; Bhattacharya, M. Elemental status in yak hair. *Indian Vet. J.* **2005**, *82*, 526–528.
57. Patrashkov, S.A.; Petukhov, V.L.; Korotkevich, O.S.; Petukhov, I.V. Content of heavy metals in the hair. *J. Phys. IV Fr.* **2003**, *107*, 1025–1027. [CrossRef]
58. Friedrichs, K.R.; Harr, K.E.; Freeman, K.P.; Szladovits, B.; Walton, R.M.; Barnhart, K.F.; Blanco-Chavez, J. ASVCP reference interval guidelines: Determination of de novo reference intervals in veterinary species and other related topics. *Vet. Clin. Pathol.* **2012**, *41*, 441–453. [CrossRef]
59. Egeland, G.M.; Ponce, R.; Bloom, N.S.; Knecht, R.; Loring, S.; Middaugh, J.P. Hair methylmercury levels of mummies of the Aleutian Islands, Alaska. *Environ. Res.* **2009**, *109*, 281–286. [CrossRef]
60. Arnold, S.; Parker, J.; Bloom, N.; Aufderheide, A.; Middaugh, J. Mercury in Ancient Mummy Hair from Peru, Chile, and Egypt—Evidence of Pre-Industrial Naturally Occurring Dietary Exposure 22 May 2018. 2018. Available online: http://dhss.alaska.gov/dph/Epi/eph/Documents/biom/Mummy%20Mercury%20Study.2018.pdf (accessed on 29 September 2020).
61. Dietz, R.; Outridge, P.M.; Hobson, K.A. Anthropogenic contributions to mercury levels in present-day Arctic animals—A review. *Sci. Total Environ.* **2009**, *407*, 6120–6131. [CrossRef]
62. Dunlap, K.L.; Reynolds, A.J.; Bowers, P.M.; Duffy, L.K. Hair analysis in sled dogs (*Canis lupus familiaris*) illustrates a linkage of mercury exposure along the Yukon River with human subsistence food systems. *Sci. Total Environ.* **2007**, *385*, 80–85. [CrossRef] [PubMed]
63. Gustin, M.S.; Bank, M.S.; Bishop, K.; Bowman, K.; Branfireun, B.; Chételat, J.; Eckley, C.S.; Hammerschmidt, C.R.; Lamborg, C.; Lyman, S.; et al. Mercury biogeochemical cycling: A synthesis of recent scientific advances. *Sci. Total Environ.* **2020**, *737*, 139619. [CrossRef] [PubMed]
64. Cooke, C.A.; Martinez-Cortizas, A.; Bindler, R.; Gustin, M.S. Environmental archives of atmospheric Hg deposition—A review. *Sci. Total Environ.* **2020**, *709*. [CrossRef] [PubMed]
65. Smulsky, J. New results on the earth insolation and their correlation with the late pleistocene paleoclimate of West Siberia. *Geol. Geophys.* **2016**, *57*, 1393–1407. [CrossRef]
66. Schuster, P.F.; Schaefer, K.M.; Aiken, G.R.; Antweiler, R.C.; Dewild, J.F.; Gryziec, J.D.; Gusmeroli, A.; Hugelius, G.; Jafarov, E.; Krabbenhoft, D.P.; et al. Permafrost Stores a Globally Significant Amount of Mercury. *Geophys. Res. Lett.* **2018**, *45*, 1463–1471. [CrossRef]
67. Jitaru, P.; Gabrielli, P.; Marteel, A.; Plane, J.M.C.; Planchon, F.A.M.; Gauchard, P.-A.; Ferrari, C.P.; Boutron, C.F.; Adams, F.C.; Hong, S.; et al. Atmospheric depletion of mercury over Antarctica during glacial periods. *Nat. Geosci.* **2009**, *2*, 505–508. [CrossRef]
68. Lambert, F.; Delmonte, B.; Petit, J.R.; Bigler, M.; Kaufmann, P.R.; Hutterli, M.A.; Stocker, T.F.; Ruth, U.; Steffensen, J.P.; Maggi, V. Dust-climate couplings over the past 800,000 years from the EPICA Dome C ice core. *Nature* **2008**, *452*, 616–619. [CrossRef]
69. Ruth, U.; Bigler, M.; Röthlisberger, R.; Siggaard-Andersen, M.-L.; Kipfstuhl, S.; Goto-Azuma, K.; Hansson, M.E.; Johnsen, S.J.; Lu, H.; Steffensen, J.P. Ice core evidence for a very tight link between North Atlantic and east Asian glacial climate. *Geophys. Res. Lett.* **2007**, *34*. [CrossRef]

70. Waelbroeck, C.; Lougheed, B.C.; Riveiros, N.V.; Missiaen, L.; Pedro, J.; Dokken, T.; Hajdas, I.; Wacker, L.; Abbott, P.; Dumoulin, J.-P.; et al. Consistently dated Atlantic sediment cores over the last 40 thousand years. *Sci. Data* **2019**, *6*. [CrossRef]
71. Kirillova, I.V.; Argant, J.; Lapteva, E.G.; Korona, O.M.; van der Plicht, J.; Zinovyev, E.V.; Kotov, A.A.; Chernova, O.F.; Fadeeva, E.O.; Baturina, O.A.; et al. The diet and environment of mammoths in North-East Russia reconstructed from the contents of their feces. *Quat. Int.* **2016**, *406*, 147–161. [CrossRef]

applied sciences

MDPI

Article

Mercury in Marine Mussels from the St. Lawrence Estuary and Gulf (Canada): A Mussel Watch Survey Revisited after 40 Years

Daniel Cossa [1,*] and Anne-Marie Tabard [2]

1 ISTerre, Université Grenoble Alpes, CS 40700, F-38058 Grenoble CEDEX 9, France
2 Eco-éducation, 4300 rue Saint André, Montréal, QC H2J 2Z2, Canada; annie.tabard@gmail.com
* Correspondence: dcossa@ifremer.fr; Tel.: +33-438-939-6907

Received: 29 September 2020; Accepted: 22 October 2020; Published: 27 October 2020

Abstract: Various species of marine mussels have been used, in the last 50 years, as sentinel organisms for monitoring metal contamination along marine coasts. There are two main reasons for this: these mollusks concentrate metals in their soft tissue and they are geographically widespread. In practice, trace metal concentrations in mussel soft tissue reveal (after some correction for biotic effects) the contamination level of their surrounding environment. We present the results of a mercury (Hg) survey in *Mytilus* spp. collected in the summers of 2016, 2018, and 2019 at 51 stations distributed along the coasts of the Estuary and Gulf of St. Lawrence. Mercury concentrations ranged from 0.063 to 0.507 $\mu g\ g^{-1}$ (dry weight, dw), with a grand mean of 0.173 ± 0.076 $\mu g\ g^{-1}$ dw (±1 standard deviation), and a median of 0.156 $\mu g\ g^{-1}$ dw for the 504 individuals analyzed. Mercury contents per individual mussel were significantly ($p < 0.01$) related to shell length and dry tissue weight, with the smaller individuals having the highest Hg concentrations. To take into account these biotic effects, we normalized Hg concentrations of the mussel soft tissue for constant shell length (L) and soft tissue weight (TW) based on the log-log relationships between Hg content and L or TW. The normalized Hg contents of mussels varied from 10.9 to 66.6 ng per virtual individual of 35 mm length and 0.17 g dry weight. A similar normalization procedure applied to 1977–1979 data, yielded a very similar range: 12 to 64 ng. This observation suggests that the Hg bioavailable to marine mussels in the study area did not change over a span of 40 years. Regional Hg distribution patterns indicate a gradual decrease of Hg content in mussels downstream from freshwater discharges to the St. Lawrence Estuary and the Baie des Chaleurs, suggesting that rivers constitute a significant Hg source in these estuarine systems. Atmospheric Hg deposition and concentration in marine waters of the Atlantic Ocean are known to have decreased in the last decades. However, in coastal environments, the response to these changes does not seem to be rapid, probably because of the long residence time of Hg in soils before being exported to coastal areas.

Keywords: mercury; mussel; mussel watch; *Mytilus*; St. Lawrence

1. Introduction

Marine mussels (*Mytilus* spp.) have been successfully used over the past fifty years as sentinel organisms for monitoring metal contamination along marine coasts [1–9]. The reason for this is that this mollusk genus is geographically widespread in sub-boreal and temperate environments and it concentrates metals in its soft tissue in proportion with the concentration in surrounding waters [10–12]. Thus, trace metal concentrations in the soft tissue of the blue mussel reveal the contamination level of the waters of its environment. Applied to the monitoring of temporal and geographical trends of chemical contamination, this approach has been named "Mussel Watch", and has been adopted as one of several coastal environmental quality-monitoring strategies by United Nations programs [3,13].

However, biological factors related to mollusk growth rate also control metal uptake and excretion and must be taken into account in order to optimize the use of mussels as sentinel organisms [4,10,14–16]. Several Mussel Watch programs have been carried out to monitor trace metal contamination along the eastern coasts of Canada and US [8,17–19], and especially mercury (Hg) contamination in the Estuary and Gulf of St. Lawrence [1,20,21]. These programs have not been maintained over more than a few years and, consequently, temporal trends are not as well documented. Nevertheless, they constitute baselines against which future assessments can be compared [22].

Here, we present the results of a Hg survey carried out in the summers of 2016, 2018, and 2019 in the St. Lawrence Estuary and Gulf, forty years after the first Mussel Watch was completed in this area. We used a normalization model to minimize the effects of biological factors on the Hg content of the mussel soft tissues. Our observations suggest that the amount of Hg bioavailable to marine mussels in the study area is similar to what it was 40 years ago. Sub-regional Hg distribution patterns indicate a gradual decrease of Hg content in mussels downstream from the main freshwater tributary (St. Lawrence River) to the St. Lawrence Estuary.

2. Materials and Methods

2.1. Sampling and Pre-Treatment

Adult specimens of *Mytilus edulis* ranging from 23 to 84 mm in length were sampled at 51 sites, at mid-tide level along the shores of the Estuary and Gulf of St. Lawrence (Figure 1). Geographical coordinates are given in Appendix A Table A1. Ten to twelve specimens (except for Station 36 with only 4 individuals) were collected at each site between 2–10 August 2016 along the South Shore of the Lower St. Lawrence Estuary (LSLE) and around the Gaspé Peninsula, between 27 August and 8 September 2018 along the North shore of the LSLE and Moyenne Côte Nord, and between 22–30 August 2019 along the Baie des Chaleurs, Northumberland Strait and Cape Breton (Figure 1). A few individuals with a shell deformity or green soft tissue were excluded to avoid the species *Coccomyxa*-infested *Mytilus trossulus*, for which background information about their trace element bioaccumulation properties is scarce [23–25]. In the field, mussel shell lengths were measured with a Vernier caliper, weighted, and their soft tissues freed from the shell. Soft tissues were kept at +4 °C in 2 mL of an ethanolic solution (90%, *v*/*v*) during the sampling periods, then kept at −18 °C until freeze-dried.

2.2. Chemical Analyses

The freeze-dried soft tissue of each mussel was individually analyzed using an automated atomic absorption spectrometer designed for Hg determinations (AMA-254, Altec, Czech Republic). The analytical procedure is described as EPA Method 7473 [26]. The analytical accuracy was checked every 10 determinations, using Certified Reference Materials (CRM): BCR-278R from the Institute for Reference Materials and Measurements, or IAEA-336 from the International Atomic Energy Agency, or DORM-4 from the National Research Council of Canada. BCR-278R is a mussel tissue, whereas IAEA-336 and DORM-4 are lichen and fish muscle, respectively. Analytical results were always within the target range of the certified values. Reproducibility, calculated as the variation coefficient (i.e., confidence interval/mean) of 6 replicate analyses of CRMs, varied between 2% and 5%. The detection limit, defined as 3 times the standard deviation of the 6 blank replicates, was 0.002 $\mu g\ g^{-1}$ (dry weight, dw). The amount of Hg dissolved in the ethanolic solution was measured in 20 samples taken at random; it never exceeded 2% of the total Hg burden in the soft tissues.

2.3. Normalization Procedure and Statistics

Mercury mussel content depends upon mussel shell length and soft tissue mass. The former because Hg accumulated throughout the lifespan of the mussel, the latter mainly because of its seasonal weight gains and losses [4,10]. To minimize the influence of mussel size on their Hg content, the normalization procedure described by Cossa and Rondeau [1] was applied. Normalization consists of correcting Hg raw data based on a multilinear regression function relating logHg content and shell length (logL) and soft tissue dry weight (logTW) (see Section 4.3). This normalization minimizes the influence of size on Hg content of the animals, thereby enabling the interpretation of mussel Hg content in terms of Hg bioavailability in their surrounding environment. According to Cossa and Rondeau [1] a two-fold Hg content distinction, environmentally-sound in terms of Hg bioavailability, can thereby be achieved.

The statistical analyses were performed with XLSTAT software from Addinsoft (https://www.xlstat.com/). As their variance tends to increase with their mean, the Hg concentration values were $\log_{10}$-transformed before any statistical treatment.

3. Hydrological and Ecological Settings of the Area of Study

The studied area comprises two main regions, the LSLE and Gulf of St. Lawrence (Figure 1). The LSLE stretches from the mouth of the Saguenay Fjord (Station 26) to Pointe des Monts (Station 35); the south and north shores of the LSLE strongly differ in their hydrographic characteristics. The south shore is characterized by an estuarine circulation as the brackish waters, flowing out the St. Lawrence Estuary, run along the south shore to form the Gaspé Current, whereas the north shore of the LSLE is affected by upwellings that bring deep waters of the Gulf of St. Lawrence to the surface. The Gulf is an epicontinental sea open to North Atlantic water inputs by two straits: the Cabot Strait between Cape Breton and Newfoundland, and the Strait of Belle Isle between Labrador and Newfoundland. The Northeastern shore of the Gulf, called "Moyenne Côte Nord" (Figure 1), is under the hydrographical influences of upwellings (especially near the Anticosti Gyre), freshwater inputs from large rivers draining the Canadian Shield, and Labrador surface waters inflowing the Gulf westward through the Strait of Belle Isle. The estuarine waters of the LSLE that spread along the Gaspé Peninsula are diluted by more salty waters from the Anticosti Gyre (Figure 1). The shores of the Baie des Chaleurs, the Northumberland Strait, and the west coast of Cape Breton are characterized by relatively shallow waters and warmer water temperatures than in the northern part of the Gulf. In addition, the surface waters of the Baie des Chaleurs are impacted by river inputs, mainly from the Restigouche River, located at the western end of the Bay. A geographical partition of the Estuary and Gulf of St. Lawrence has been proposed based on ecological criteria [27]. These authors concluded that the most biologically significant hydrological features of the system are the LSLE, the Gaspé Current that hugs the coast of the Gaspé Peninsula, and the Northwestern Gulf, i.e., the Anticosti Gyre whose waters intercept the Moyenne Côte Nord west of Anticosti Island. These findings are supported by fluorescence distributions in the surface waters of the St. Lawrence Estuary and Gulf [28]. Based on the above-listed hydrological and ecological criteria, we decided to distinguish the following six regions: (1) North shore of the LSLE, (2) Moyenne Côte Nord, (3) South shore of the LSLE, (4) Gaspé Peninsula, and (5) Baie des Chaleurs, and (6) Northumberland-Cape Breton continuum.

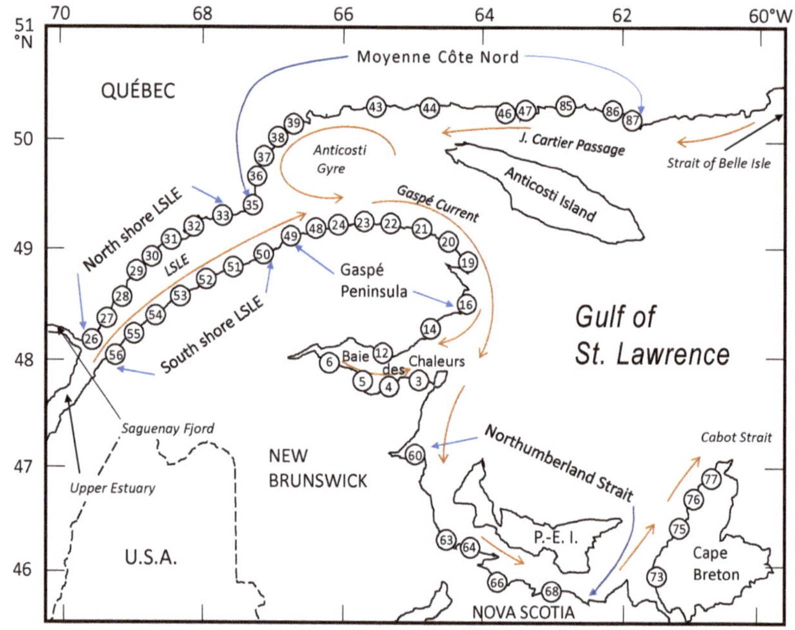

Figure 1. Location of the sampling sites and hydro-ecological regions. LSLE: Lower St. Lawrence Estuary. P.-E.I. Prince Edward Island. Brown arrows indicate main surface water circulation.

4. Results and Discussion

4.1. Geographical Patterns of the Growth

Total shell length (L), width (l), height (h), and tissue mass (M) measurements on mussel allow us to calculate allometric growth indices, that may vary depending on the *Mytilus* species and growth conditions [29,30]. In the St. Lawrence Estuary and Gulf, L, l, and h are linearly correlated, whereas L and M are better related by power functions [17]. These findings are confirmed by the present data (Table 1, Figure 2). In this study (2016–2019), allometric growth indices (h/L, l/L ratios, and the parameters of the L vs M power functions) were similar to the earlier sampling (1977–1979) (Table 1). This similarity in the mussel morphometry suggests that the possible environmental changes in the St Lawrence Estuary and Gulf [31] during the last decades, such as temperature, did not generate shell shape differences despite the high plasticity of the mussel shell [32]. Nevertheless, the parameters of the L vs M equations, derived for each of the six regions, display some variations (Figure 2, Table 1). Such slight regional morphological shell differences are consistent with the small regional differences in the absolute growth rate estimated in 1977–1979 [17]. We can infer from these allometric index comparisons that mussel growth conditions have not differed significantly for the last forty years, with slight regional differences in the L *vs* M index persisting (Table 1). This finding allows us to compare the Hg load of the mussel soft tissues between the two time periods without any significant bias affecting Hg bioaccumulation due to the change of mussel growth rates.

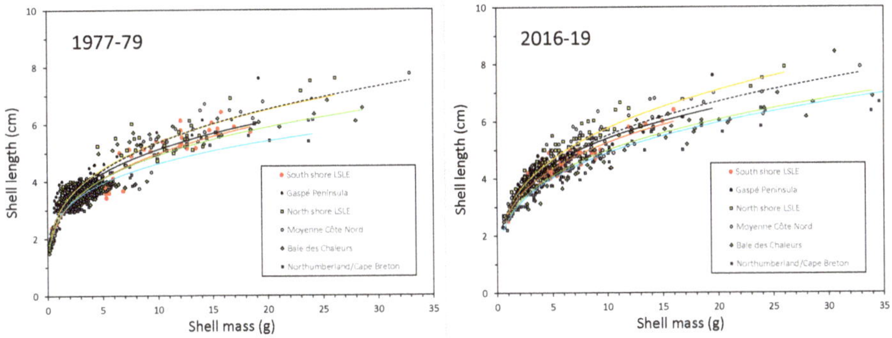

Figure 2. Length-mass relationship of *Mytilus edulis* shell from six hydro-ecological regions of the St. Lawrence Estuary and Gulf (Figure 1). Mussels sampled in 1977–1979 (n = 1222) and in 2016–2019 (n = 504) periods.

Table 1. Allometric relationships in the mussel shell shape and their changes between 1977–1979 and 2016–2019 periods. L: length (cm), l: width (cm); h: height (cm); M: mass (g). All the relationships were statistically significant with a $p < 0.001$. AGR: Absolute growth rate (cm y^{-1}) from ref. [17]. 1977–1979 statistics were calculated on 1222 individual mussels vs. 504 in 2016–2019.

Period	h/L	l/L	L vs M	AGR
		South shore LSLE		
1977–1979	0.52 ± 0.03	0.40 ± 0.07	$L = 2.62\ M^{0.28}$	0.57
2016–2019	0.49 ± 0.03	0.42 ± 0.04	$L = 2.92\ M^{0.25}$	–
		Gaspé Peninsula		
1977–1979	0.50 ± 0.05	0.42 ± 0.03	$L = 2.85\ M^{0.25}$	–
2016–2019	0.49 ± 0.05	0.41 ± 0.04	$L = 2.97\ M^{0.26}$	–
		North shore LSLE		
1977–1979	0.50 ± 0.05	0.42 ± 0.03	$L = 2.85\ M^{0.25}$	0.52–0.60
2016–2019	0.51 ± 0.03	0.42 ± 0.07	$L = 2.88\ M^{0.29}$	
		Moyenne Côte Nord		
1977–1979	0.52 ± 0.04	0.42 ± 0.05	$L = 2.78\ M^{0.28}$	0.57–0.65
2016–2019	0.52 ± 0.04	0.42 ± 0.03	$L = 2.89\ M^{0.28}$	–
		Baie des Chaleurs		
1977–1979	0.52 ± 0.04	0.42 ± 0.05	$L = 2.67\ M^{0.26}$	0.56–0.67
2016–2019	0.56 ± 0.04	0.43 ± 0.04	$L = 2.60\ M^{0.28}$	–
		Northumberland/Cape Breton		
1977–1979	0.57 ± 0.06	0.42 ± 0.05	$L = 2.72\ M^{0.23}$	0.40–0.57
2016–2019	0.57 ± 0.06	0.42 ± 0.04	$L = 2.54\ M^{0.28}$	–
		Estuary and Gulf of St Lawrence		
1977–1979	0.53 ± 0.06	0.43 ± 0.06	$L = 2.80\ M^{0.26}$	0.40–0.57
2016–2019	0.52 ± 0.07	0.43 ± 0.04	$L = 2.83\ M^{0.27}$	–

4.2. Estuary and Gulf of St. Lawrence Mussel Watch in the North Atlantic Context

The Hg concentrations in mussel soft tissues ranged from 0.06 to 0.51 µg g^{-1} dw (n = 504), with a grand mean of 0.17 ± 0.08 µg g^{-1} dw (± 1 standard deviation), and a median of 0.16 µg g^{-1} dw. In 1977–1979, the Hg concentrations were very similar with a mean of 0.16 ± 0.05 µg g^{-1} dw, and a median of 0.15 µg g^{-1} dw (n = 442) [33]. Very few Mussel Watch programs have been carried out uninterrupted for longer than a few years with the notable exception of the US and French monitoring programs that have been running yearly since the 1980s [34,35]. The median concentration in the St. Lawrence system is slightly higher than the most recent available medians: 0.11 µg g^{-1} dw (2005–2012, n = 298) for the US Mussel Watch and 0.12 µg g^{-1} dw (2000–2004, n = 303) for the French Mussel Watch programs [34,35]. It is similar to the annual median concentrations of the French coasts

Mussel Watch for the period 1980–94 (0.15 μg g^{-1} dw, n = 937) [36,37]. In summary, the current median Hg concentrations of the St. Lawrence Mussel Watch is similar to the long-term representative surveys carried out along the Atlantic shores in the last 40 years.

4.3. Temporal and Regional Hg Trends

To reach an optimal discriminating capacity for detecting temporal and regional trends, we used a normalization procedure minimizing bias due to biological factors (see above). The relationships between the Hg content of the 504 analyzed mussels and the dry mass of their soft tissue (TW) or shell length (L) were highly significant ($p < 0.001$). The Equation used for the normalization model built with the entire sampling set is:

$$\log Hg = 1.39 \pm 0.15 \log L + 0.49 \pm 0.04 \log TW - 0.36 \ (R^2 = 0.87; n = 504) \quad (1)$$

compared to:

$$\log Hg = 1.43 \pm 0.49 \log L + 0.34 \pm 0.16 \log TW - 0.48 \ (R^2 = 0.91; n = 143) \quad (2)$$

obtained for the St Lawrence Mussel Watch performed in 1977–1979 [1]. Regression coefficients, obtained forty years apart, are not statistically different ($p < 0.01$).

Normalized Hg content (individual of 35 mm length and 0.17 g dry weight) varied from 10.9 to 66.6 ng per individual, with a grand mean of 42.0 ± 2.1 ng (±1 standard deviation), and median of 44.2 ng, for the 504 individuals. Expressed as Hg concentration in the soft tissue, the range becomes 0.06–0.39 μg g^{-1} (dw). A similar normalization procedure applied to the 1977–1979 data, gave a very similar range: 12 to 64 ng per individual mussel [1] for the same stations. This finding strongly suggests that the Hg bioavailable to marine mussels in the study area has not changed over 40 years. At first glance, these observations are surprising since a decrease in Hg inputs to the St. Lawrence Estuary and Gulf waters could be expected (see below).

First, riverine Hg inputs to the St. Lawrence Estuary should have decreased due to the implementation of the International regulation on the Great Lakes basin [38–40]. Few time-series observations of Hg inputs from riverine sources, but a high-frequency sampling experiment of water was performed over 18 months in 1995–96 in the St. Lawrence River. Mercury export to the LSLE was estimated at ~1.2 Mg y^{-1}, for a mean dissolved Hg concentration of 0.60 ± 0.46 ng L^{-1} [41]. Since that period, no systematic study has been published on temporal variations of Hg concentrations in the St. Lawrence River. According to a recent government report, the Hg flux to the St. Lawrence River has not changed between 1995–1996 and 2004–2008 [42]. A comparison of mussel mean Hg contents at stations from the head of the LSLE (Stations 53–56), where freshwater influence is maximum, fails to show a statistically significant difference ($p < 0.01$) in Hg tissue levels between 1977–1979 (44.6 ± 6.9 ng, n = 5) and 2016–2019 (37.4 ± 6.6 ng, n = 5). It could be interpreted to imply that Hg bioavailability in freshwaters of the LSLE has not changed in the last 40 years.

Secondly, atmospheric Hg concentrations and wet deposition in Eastern North America are thought to have declined during the 1990–2010 period [43–45]. Regulations for reducing Hg emissions were implemented in New England and Eastern Canada [46]. In an assessment of Hg sources and fate in a marine environment very close to the Gulf of St Lawrence (The Gulf of Maine), Sunderland et al. [47] reported that: "Temporal patterns in sentinel species (mussels and birds) have in some cases declined in response to localized point source mercury reductions but overall Hg trends do not show consistent declines". Likewise, studies of coastal Massachusetts, New Hampshire, and Maine, reveal that no significant temporal trends in mussel Hg concentrations were found between 1990 and 2010 at 12 of the 15 monitored stations [19]. Our results are also consistent with those of Hg trends in herring gull eggs from Atlantic Canada collected between 1972 and 2008 [48]. These authors reported that, after adjusting Hg trends for dietary shifts, environmental Hg in coastal ecosystems had remained relatively constant in Eastern Canada over the previous 36 years. According to Sunderland et al. [43], reductions in atmospheric Hg deposition from North American sources could have been offset by increased deposition from global Hg sources.

Figure 3 illustrates the station-to-station variability in mussel tissue concentration. Strikingly, the continuum on the South shore of the LSLE- Gaspé Peninsula (Stations 56 to 14) exhibits a gradual Hg decrease from the brackish estuarine water originating from the St Lawrence Estuary to the Gaspesian coast as they are diluted by the Gulf waters originating from the Anticosti Gyre. This pattern strongly suggests that St. Lawrence River waters are a significant source of bioavailable Hg. Seaward, along the Gaspé Current, mussels exhibit very low Hg contents and variability (Stations 48, 49, 16 to 24, Figures 1 and 3). This would imply that Gaspé Current waters are impoverished in bioavailable Hg compared to waters of the LSLE. Increasing Hg reduction and evasion in the atmosphere in the productive waters (Anticosti Gyre and Gaspé Current, see Section 3) may favor this Hg depletion. It should be noted that the Gaspesian coastline is also an area where riverine inputs are small compared to other LSLE and Gulf shores. Another Hg dilution structure is visible along the Baie des Chaleurs (Stations 6 to 3, Figures 1 and 3) where riverine inputs are also important (see Section 3). Interesting to note is that, during the 1077–79 Mussel Watch, high Hg contents in mussel soft tissue were also observed at stations along the southern shores of the LSLE and of the Baie des Chaleurs where brackish waters are present (see Figure 4 in Ref. [1]).

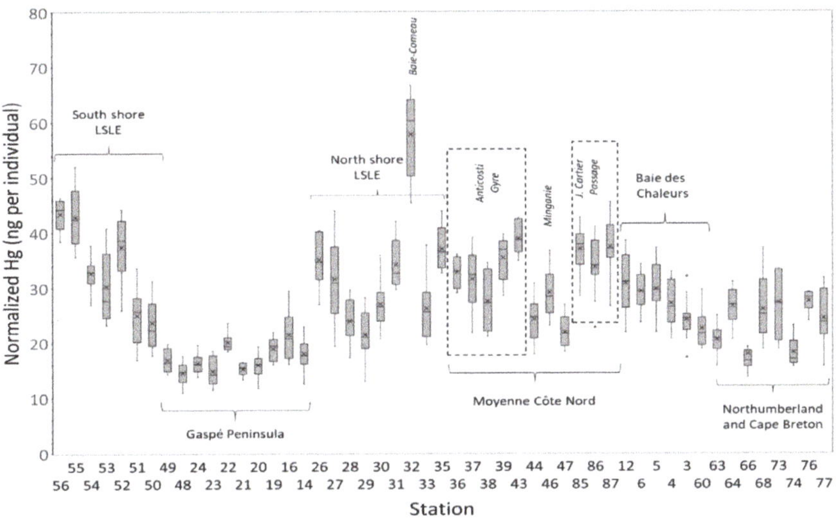

Figure 3. Mercury content (ng) of individual mussels after normalization to 35 mm-long and 0.17 g virtual specimens. Mean contents are indicated by a horizontal bar and the median by a cross. Each station combines 10 to 12 individual mussels. Stations are eastward distributed on the x-axis.

The North Shore of the LSLE and the Moyenne Côte Nord coastlines display inter-station large variations in mussel tissue Hg content, even within the same oceanographic entity (e.g., Moyenne Côte Nord, Figures 1 and 3). These fluctuating distributions are difficult to interpret since no steady geographical trend can be observed in Hg content. Nevertheless, some hypotheses can be put forward to explain these. Mercury content in the coastal waters of the North Shore of the St. Lawrence Estuary and Gulf are most likely influenced by upwellings and organic-rich river waters that drain through the Canadian Shield soils [49,50]. The organic matter of these circumneutral pH waters contains strong Hg-binding functional groups (e.g., thiol) that favor Hg transport in solution [41,51–53]. On the other hand, upwellings bring saline low Hg waters of Atlantic origin to the surface, especially in the Minganie region (Stations 44–47, Figures 1 and 3) [54]. These Atlantic waters enter the Gulf through

Cabot Strait and are known to contain low picomolar Hg levels [55]. Conversely, waters entering the Northeastern Gulf through the Strait of Belle Isle originate from the Labrador Sea Current, known for its relative high Hg concentrations due to the inputs of the organic-rich rivers outflowing the Arctic Canadian Archipelago [56]. These waters, brought by the Labrador Sea Current, are transported westward along the J. Cartier Passage, a remote region without any known trace-metal sources but with relatively high Hg content in sampled mussels (Figures 1 and 3).

It is interesting to note that low Hg content mussels are located along the Gaspé Peninsula (Stations 48, 49, 16–24) and Minganie (Stations 44–47) coasts, where commercial or experimental mussel farms are installed (mapaq.gouv.qc.ca/fr/Peche/aquaculture) [57,58]. This latter area is known for its low-temperature upwellings in summer [54].

4.4. Current and Former Hot Spots

Industrial areas distributed along the Estuary and Gulf coastlines are likely to shelter polluted sites. These would include a former chlor-alkali plant, pulp and paper mills, and smelters [33]. The highest Hg concentrations of the present survey were found at Station 32 near the city of Baie Comeau. The city hosts an active industrial complex and the bay sediments are enriched in trace metals from current and legacy sources [59,60]. Our study fails to highlight other potential hotspots that had been identified by previous Mussel Watch Programs. A chlor-alkali plant located near the mouth of the Restigouche River in the Baie des Chaleurs has been shown to generate local Zn and Hg contamination [61], but its impact on mussel tissue Hg contents was not revealed by our study. This might be explained by may be the result of the 20-km distance between Station 6 and the plant and/or the fact that it was shut down in 2008 [62]. Likewise, the previously reported Hg contamination [20] at the mouth of the Saguenay Fjord (Station 26), is no longer being observed, as the main Hg source, the Arvida chlor-alkali plant was shut down 40 years ago [63].

5. Summary and Conclusions

A Hg Mussel Watch Program was carried out at 52 stations distributed along the intertidal zone of the Estuary and Gulf of St. Lawrence, in the summers of 2016, 2018, and 2019. This survey took place forty years after a previous Mussel Watch survey was performed in this area (1977–1979), at almost all stations currently sampled. The same normalization procedure applied 40 years ago was also applied to the current data set. It allows the minimization of the effects of biological factors on the Hg content of the mussel soft tissues and optimizes the interpretation of Hg distribution in terms of the availability of Hg in coastal waters. Current Hg concentrations in the soft tissue of mussels from the St. Lawrence System are similar to those of the perennial Mussel Watch Programs implemented along the US and French Atlantic coasts [34,35]. The results indicate that the normalized Hg contents of mussels in the Estuary and the Gulf of St Lawrence were similar at sampling times 40 years apart. In addition, sub-regional Hg distribution patterns indicate a gradual decrease of Hg content in mussels downstream from freshwater inputs, which suggests that the spatial distribution of Hg concentrations in the soft tissues of marine mussels is, in part, governed by riverine Hg sources. Atmospheric Hg deposition and concentration in marine waters of the Atlantic Ocean are known to have decreased in the last decades [44,64]. However, in coastal environments, the response to these changes does not seem to be rapid. This probably results from the long residence time of Hg in soil and land cover before being exported to coastal areas with freshwater discharges. Indeed, different environmental reservoirs have different time scales as regards Hg mobility [65,66]. Furthermore, coastal sediment resuspension/sedimentation cycling may also contribute to a longer retention time of Hg in the intertidal marine environments, making mussel habitats conducive to the retention of legacy Hg. The changes in atmospheric Hg inputs are thus damped in terms of Hg availability for sessile benthic animals. A renewed Mussel Watch survey in a few decades may elucidate the long-time trend of Hg load in the St Lawrence Coastal System.

Author Contributions: D.C. designed the experiment, performed sampling and analyses, and wrote the article. A.-M.T. participated in sampling and sample preparation. All authors have read and agreed to the published version of the manuscript.

Funding: This research received no external funding.

Acknowledgments: Thanks are due to A. Mucci for providing some of the sampling material and S. Guédron for his support in mercury analyses performed within the analytical chemistry platform of ISTerre (OSUG-France). Special thanks to E. Bourget, A. Mucci, J.-M. Sévigny, and M. Vautour for their comments on the manuscript.

Conflicts of Interest: The authors declare that there is no conflict of interest regarding the publication of this paper.

Appendix A

Table A1. Station coordinates. See also map in Figure 1. Mean normalized Hg content of (±1 SD) in ng per individual.

Station	Name	Longitude (W)	Latitude (N)	Hg Content
	South shore of the LSLE/Gaspé Peninsula			
56	Rivière-Trois-Pistoles	69°13′18.1″	48°06′22.3″	43.4 ± 2.8
55	St-Simon-sur-Mer	69°03′39.4″	48°13′10.5″	41.9 ± 5.9
54	Cap du Corbeau (Bic)	68°43′38.2″	48°22′44.6″	32.7 ± 3.1
53	Anse aux coques (St Luce)	68°22′55.8″	48°33′05.3″	30.3 ± 6.4
52	Pointe-Mitis	68°01′55.5″	48°40′52.4″	37.4 ± 6.0
51	Petite rivière Blanche (St Ulric)	67°41′31.6″	48°47′36.5″	25.0 ± 5.2
50	Anse à la Croix	67°18′00.7″	48°54′27.1″	23.7 ± 4.7
49	Baie des Capucins	66°50′37.8″	49°03′09.4″	16.8 ± 2.2
48	Anse de l'Eglise (Tourelle)	66°25′12.6″	49°09′27.6″	14.5 ± 2.2
24	Cap à la Martre	66°10′19.0″	49°12′26.7″	16.3 ± 1.8
23	Anse de Mont-St-Pierre	65°47′32.0″	49°13′40.0″	14.8 ± 2.5
22	Rivière-la-Madeleine	65°17′45.0″	49°14′30.0″	20.1 ± 1.9
21	Pointe-à-la-Frégate	64°56′25.6″	49°12′28.0″	15.2 ± 1.1
20	Pointe Jaune	64°30′17.3″	49°04′07.2″	16.0 ± 2.2
19	Cap des Rosiers	64°12′08.0″	48°51′21.0″	18.9 ± 2.2
16	Rocher Percé	64°12′27.7″	48°31′30.8″	21.5 ± 4.5
	North shore of the LSLE			
26	Pointe aux Vaches (Tadoussac)	69°41′18.9″	48°08′08.9″	35.0 ± 4.9
27	Grandes-Bergeronnes (Marina)	69°33′09.9″	48°13′42.9″	31.5 ± 7.9
28	Baie St Onge (Les Escoumins)	69°23′48.5″	48°21′03.2″	23.9 ± 3.8
29	Pointe Verte (Forestville)	69°03′13.4″	48°44′21.9″	21.5 ± 4.6
30	Anse du Cap Colombier	68°52′50.6″	48°49′23.5″	27.0 ± 4.9
31	Baie Verte (Ragueneau)	68°34′33.6″	49°03′38.7″	34.2 ± 4.8
32	Baie-Comeau	68°08′25.6″	49°13′12.2″	57.8 ± 8.1
33	Franquelin	67°53′46.0″	49°17′28.7″	26.2 ± 6.0
	Moyenne Côte Nord			
35	Pointe des Monts	67°22′14.6″	49°19′30.5″	37.0 ± 4.3
36	Pointe aux Anglais	67°12′09.0″	49°38′14.1″	32.8 ± 3.0
37	Anse au Pot	67°08′57.1″	49°48′04.0″	31.5 ± 5.4
38	La Grande Anse	66°56′56.1″	49°59′00.3″	27.5 ± 5.8
39	Parc Île Paterson (Port Cartier)	66°52′12.8″	50°01′05.3″	35.4 ± 4.0
43	Rivière Pigou	65°30′37.9″	50°16′59.6″	38.8 ± 3.0
44	Rivière-au-Tonnerre	64°45′42.0″	50°16′22.4″	24.2 ± 3.9
46	Île Quarry	63°50′11.8″	50°12′28.3″	29.1 ± 4.4
47	Havre-St-Pierre	63°36′08.5″	50°14′12.5″	21.9 ± 2.7
85	Baie Johan Beetz	62°48′26.9″	50°17′10.1″	37.0 ± 3.9
86	Île Michon	62°01′53.9″	50°13′21.8″	33.7 ± 5.2
87	Le Petit Havre (Natashquan)	61°50′26.6″	50°11′19.9″	37.4 ± 5.4

Table A1. *Cont.*

Station	Name	Longitude (W)	Latitude (N)	Hg Content
		Baie des Chaleurs		
14	Point Newport	64°43′42.6″	48°17′08.6″	18.0 ± 3.0
12	Pointe Bonaventure	65°27′33.1″	48°00′25.8″	30.8 ± 5.8
6	Nash Creek	66°07′48.7″	45°55′08.3″	29.3 ± 3.2
5	Petit-Rocher	65°42′37.0″	47°47′02.6″	29.7 ± 4.2
4	Stonehaven	65°21′43.5″	47°45′14.6″	27.1 ± 4.4
3	Anse-Bleue	65°06′13.4″	47°49′46.7″	24.2 ± 3.9
		Northumberland/Cape Breton		
60	Escuminac	64°54′47.2″	47°04′39.7″	22.5 ± 3.8
63	Cap Bimet	64°27′31.9″	46°14′14.8″	20.6 ± 2.8
64	Cap Pelé (Trois-Ruisseaux)	64°13′35.2″	46°13′16.6″	26.6 ± 3.3
66	Wild Rose (Linden)	63°47′02.3″	45°53′17.8″	17.9 ± 4.8
68	Horn Point (Seafoam)	63°00′30.3″	45°47′40.0″	26.1 ± 5.7
73	Little Judique Ponds	61°30′44.9″	45°54′28.3″	27.3 ± 6.1
75	Whale Cove (Margaree)	61°08′18.2″	46°25′28.2″	18.2 ± 2.6
76	Pointe Enragée (Chéticamp)	61°01′37.9″	46°38′58.2″	27.5 ± 1.9
77A	Pleasant Bay (Harbour)	60°47′52.0″	46°49′53.2″	24.4 ± 5.1
77B	MacKenzies River/Pleasant Bay	60°49′51.5″	46°49′25.0″	

References

1. Cossa, D.; Rondeau, J.G. Seasonal, geographical and size induced variability of mercury content of *Mytilus edulis* (L.) in an estuarine environment: A re-assessment of mercury pollution level in the Estuary and Gulf of St. Lawrence. *Mar. Biol.* **1985**, *88*, 43–49. [CrossRef]
2. Thibaud, Y. Teneurs en mercure dans les moules du littoral français (Mercury concentrations in marine mussels from French coastlines). *Science et Pêche (Bull. Inst. Pêches marit.)* **1973**, *221*, 1–6.
3. Goldberg, E.D. The Mussel Watch: A first step in global marine monitoring. *Mar. Pollut. Bull.* **1975**, *6*, 111–113. [CrossRef]
4. Phillips, D.J.H. *Quantitative Aquatic Biological Indicators: Their Use to Monitor Trace Metal and Organochlorine Pollution*; Applied Science Publishers Ltd.: London, UK, 1980; p. 72.
5. Farrington, J.W.; Goldberg, E.D.; Risebrough, R.W.; Martin, J.H.; Bowen, V.T. US "Mussel Watch" 1976–1978: An Overview of the Trace-Metal, DDE, PCB, hydrocarbon and Artificial Radionuclide Data. *Environ. Sci. Technol.* **1983**, *17*, 490–496. [CrossRef] [PubMed]
6. Claisse, D. Chemicals concentration of French coast: The result of ten-year mussel watch. *Mar. Pollut. Bull.* **1989**, *20*, 523–528. [CrossRef]
7. Sericano, J.L.; Wade, T.L.; Jackson, T.J.; Brooks, J.M.; Tripp, B.W.; Farrington, J.W.; Mee, L.D.; Readmann, J.W.; Villeneuve, J.P.; Goldberg, E.D. Trace organic contamination in the Americas: An overview of the US national status and trends and the international "Mussel Watch" programs. *Mar. Pollut. Bull.* **1995**, *31*, 214–225. [CrossRef]
8. Fraser, M.; Surette, C.; Vaillancourt, C. Spatial and temporal distribution of heavy metal concentrations in mussels (*Mytilus edulis*) from the Baie des Chaleurs, New Brunswick, Canada. *Mar. Pollut. Bull.* **2011**, *62*, 1345–1351. [CrossRef] [PubMed]
9. Elkus, A.A.; Leblanc, L.A.; Latimer, J.S.; Page, D.S.; Harding, G.C.H.; Wells, P.G. Monitoring chemical contaminants in the Gulf of Maine, using sediments and mussels (*Mytilus edulis*): An evaluation. *Mar. Pollut. Bull.* **2020**, *153*, 110956. [CrossRef]
10. Bayne, B. Mussel watching. *Nature* **1978**, *275*, 87–88. [CrossRef]
11. Cossa, D. A Review of the Use of *Mytilus* spp. as Quantitative Indicators of Cadmium and Mercury Contamination in Coastal Waters. *Oceanol. Acta* **1989**, *12*, 417–432.
12. Casas, S.; Gonzalez, J.-L.; Andral, B.; Cossa, D. Discriminating Physiological Influences from Element Bioavailability in Water on the Hg, Pb, Cd and Cu Content of Marine Mussels during Transplantation Experiments. *Environ. Toxico. Chem.* **2008**, *27*, 1543–1552. [CrossRef]

13. Farrington, J.H.; Tripp, B. *International Mussel Watch Project NOAA*; Technical Memorandum NOS ORCA 95; NOAA: Silver Spring, MD, USA, 1995.
14. Lobel, P.B.; Wright, D.A. Relationship between body zinc concentration and allometric growth measurements in the mussel *Mytilus edulis*. *Mar. Biol.* **1982**, *66*, 145–150. [CrossRef]
15. Riget, F.; Johansen, P.; Asmund, G. Influence of Length on Element Concentrations in Blue Mussels (*Mytilus edulis*). *Mar. Pollut. Bull.* **1996**, *32*, 745–751. [CrossRef]
16. Boeing, D.W. An evaluation of bivalves as biomonitors of heavy metals pollution in marine waters. *Environ. Monit. Assess.* **1999**, *55*, 459–470. [CrossRef]
17. Cossa, D.; Bourget, E. Croissance et morphologie de la coquille de *Mytilus edulis* L. de l'estuaire et du golfe du Saint Laurent. *Naturaliste canadien (Rev. Ecol. Syst.)* **1985**, *112*, 417–423.
18. Veinott, G.; Whalen, R.; Miller-Banoub, J. *Trace Elements and Heavy Metals Data for Blue Mussels (Mytilus edulis) from Trinity Bay and Placentia Bay, Newfoundland*; Canadian Manuscript Report of Fisheries and Aquatic Sciences No. 2643; Science, Oceans and Environment Branch, Oceans Programs Division, Department of Fisheries and Oceans: St. John's, NL, Canada, 2003; p. 9.
19. Kimbrough, K.L.; Johnson, W.E.; Lauenstein, G.G.; Christensen, J.D.; Apeti, D.A. *An Assessment of Two Decades of Contaminant Monitoring in the Nation's Coastal Zone*; NOAA Technical Memorandum NOS NCCOS 74; NOAA: Silver Spring, MD, USA, 2008; p. 105.
20. Bourget, E.; Cossa, D. Mercury content of mussels from the St. Lawrence estuary and Northwestern gulf of St. Lawrence, Canada. *Mar. Pollut. Bull.* **1976**, *7*, 237–239. [CrossRef]
21. Kennedy, K.M.; Benson, R.J. Report of Heavy Metal Analysis Conducted on Mussel (*Mytilus edulis*) Samples Collected at 55 Sites in Newfoundland. In *Canadian Technical Report of Fisheries and Aquatic Science*; Fisheries and Oceans Canada: Ottawa, ON, Canada, 1993; p. 17.
22. Farrington, J.W.; Tripp, B.W.; Tanabe, S.; Subramanian, A.; Sericano, J.L.; Wade, T.L.; Knap, A.H. Edward D. Goldberg's proposal of "the Mussel Watch": Reflections after 40 years. *Mar. Pollut. Bull.* **2016**, *110*, 501–510. [CrossRef]
23. Piwoni-Piórewicz, A.; Kukliński, P.; Strekopytov, S.; Humphreys-Williams, E.; Najorka, J.; Iglikowska, A. Size effect on the mineralogy and chemistry of *Mytilus trossulus* shells from the southern Baltic Sea: Implications for environmental monitoring. *Environ. Monit. Assess.* **2017**, *189*, 197. [CrossRef]
24. Zuykov, M.; Anderson, J.; Archambault, P.; Dufresne, F.; Pelletier, E. *Mytilus trossulus* and hybrid (*M. edulis-M. trossulus*)—New hosts organisms for pathogenic microalgae *Coccomyxa* sp. from the Estuary and northwestern Gulf of St. Lawrence, Canada. *J. Invertebr. Pathol.* **2018**, *153*, 145–146. [CrossRef]
25. Zuykov, M.; Anderson, E.; Pelletier, E. Does photosynthesis provoke formation of shell deformity in wild mytilid mussels infested with green microalgae Coccomyxa?—A conceptual model and research agenda. *J. Exp. Mar. Biol. Ecol.* **2018**, *505*, 9–11. [CrossRef]
26. EPA-Method 7473: Mercury in Solids and Solutions by Thermal Decomposition, Amalgamation, and Atomic Absorption Spectrophotometry. Available online: www.epa.gov/sites/production/files/2015-12/documents/7473.pdf (accessed on 2 September 2020).
27. Lavoie, D.; Starr, M.; Zakardjian, B.; Larouche, P. Identification of ecologically and biologically significant areas (EBSA) in the Estuary and Gulf of St. Lawrence: Primary production, 2007. Canadian Science Advisory Secretariat Research Document 2007/079; Canadian Science Advisory Secretariat. Available online: http://www.dfo-mpo.gc.ca/csas (accessed on 2 September 2020).
28. Dinauer, A.; Mucci, A. Spatial variability in surface-water pCO_2 and gas exchange in the world's largest semi-enclosed estuarine system: St Lawrence Estuary (Canada). *Biogeosciences* **2017**, *14*, 3221–3237. [CrossRef]
29. Bayne, B.L. *Marine Mussels: Their Ecology and Physiology*; International Biological Progrogramme 10; Cambridge University Press: Cambridge, UK, 1976; 506p, ISBN 978-0-521-11288-8.
30. Gardner, J.P.A.; Skibinski, D.O.F.; Bajdik, C.D. Shell growth and variability differences in the marine mussels *Mytilus edulis* (L), *Mytilus galloprovincialis* (LMK), and their hybrids from 2 sympatric populations in SW England. *Biol. Bull.* **1993**, *185*, 405–416. [CrossRef] [PubMed]
31. Dufour, R.; Benoit, H.; Castonguay, M.; Chassé, J.; Devine, L.; Galbraith, P.; Harvey, M.; Larouche, P.; Leszard, S.; Petrie, B.; et al. Ecosystem Status and Trends Report: Estuary and Gulf of St. Lawrence Ecozone. Technique Report; Canadian Science Advisory Secretariat: St. Lawrence, QC, Canada, 2010.
32. Telesca, L.; Michalek, K.; Sanders, T.; Peck, L.S.; Thyrring, J.; Harper, E.M. Blue mussel shell shape plasticity and natural environments: A quantitative approach. *Sci. Rep.* **2018**, *8*, 2865. [CrossRef] [PubMed]

33. Cossa, D. *Utilisation de la Moule Bleue Comme Indicateur du Niveau de Pollution par les Métaux Lourds et les Hydrocarbures dans L'estuaire et le Golfe du St Laurent (Using Blue Mussel as a Sentinel Organism for Monitoring Metal and Hydrocarbon Contamination in the Estuary and Gulf of St. Lawrence)*; INRS-Océanologie; Rapport de l'Institut national de la Recherche Scientifique: Rimouski, QC, Canada, 1980.
34. NCCOS/NOAA. National Centers for Coastal Ocean Science. NOAA's National Status and Trends Data. Available online: products.coastalscience.noaa.gov/nsandt_data/data.aspx (accessed on 2 September 2020).
35. RNO (Réseau National d'Observation). *Surveillance du Milieu Marin. Travaux du RNO.Ifremer et Ministère de l'Écologie et du Développement Durable*; National Observation Network, 2006 edition; French Department for Ecology and Sustainable Development: Paris, France, 2006.
36. Beliaeff, B.; O'Connor, T.P.; Claisse, D. Comparison of chemical concentrations in mussels and oysters from the United States and France. *Environ. Monit. Assess.* **1998**, *49*, 87–95. [CrossRef]
37. Claisse, D.; Cossa, D.; Bretaudeau-Sanjuan, J. Methylmercury in molluscs along the French coast. *Mar. Pollut. Bull.* **2001**, *42*, 329–332. [CrossRef]
38. Anonymous. Regulation Agencies release Great Lake protection plans. *Pollut. Eng.* **2000**, *32*, 19.
39. Edgar, S.L.; Martin, T.D. Great-Lakes protection fund. *J. Great Lakes Res.* **1988**, *14*, 381. [CrossRef]
40. Great Lakes Protection Act. Ontario Strengthens Environmental Protections for the Great Lakes New Legislation Passes to Keep the Great Lakes Clean, Swimmable and Fishable October 7, 2015. Available online: https://news.ontario.ca/en/release/34457/ontario-strengthens-environmental-protections-for-the-great-lakes (accessed on 30 August 2019).
41. Quémerais, B.; Cossa, D.; Rondeau, B.; Pham, T.T.; Gagnon, P.; Fortin, B. Sources and Fluxes of Mercury in the St. Lawrence River. *Environ. Sci. Technol.* **1999**, *33*, 840–849. [CrossRef]
42. Rondeau, M. Monitoring the state of the St. Lawrence River, Saint Lawrence Action Plan. *Environ. Can. Sci. Techol. Branch Cat* **2015**, 9, N: En154-79/2015E-PDF. Available online: planstlaurent.qc.ca/fileadmin/publications/fiches_indicateurs/Anglais/2015_Water_Quality_Fluvial.pdf (accessed on 10 August 2020).
43. Sunderland, E.M.; Cohen, M.; Selin, N.; Chmura, G. Reconciling models and measurements to assess trends in atmospheric mercury deposition. *Environ. Pollut.* **2008**, *156*, 526–535. [CrossRef] [PubMed]
44. Prestbo, E.M.; Gay, D.A. Wet deposition of mercury in the US and Canada, 1996–2005: Results and analysis of the NADP mercury deposition network (MDN). *Atmos. Environ.* **2009**, *43*, 4223–4233. [CrossRef]
45. Muntean, M.; Janssens-Maenhout, G.; Song, S.; Selin, N.E.; Olivier, J.G.J.; Guizzardi, D.; Mass, R.; Dentener, F. Trend analysis from 1970 to 2008 and model evaluation of EDGARv4 global gridded anthropogenic mercury emissions. *Sci. Total Environ.* **2014**, *494–495*, 337–350. [CrossRef] [PubMed]
46. NEG-ECP. Report to the New England Governors and Eastern Canadian Province Premiers on Mercury Projects. Available online: www.dem.ri.gov/topics/pdf/neg2003.pdf (accessed on 12 August 2020).
47. Sunderland, E.M.; Amirbahman, A.; Burgess, N.M.; Dalziel, J.; Harding, G.; Jones, S.H.; Kamai, E.; Karagas, M.R.; Shi, X.; Chen, C.Y. Mercury sources and fate in the Gulf of Maine. *Environ. Res.* **2012**, *119*, 27–41. [CrossRef] [PubMed]
48. Burgess, N.M.; Bond, A.L.; Hebert, C.E.; Neugebauer, E.; Champoux, L. Mercury trends in herring gull (*Larus argentatus*) eggs from Atlantic Canada, 1972–2008: Temporal change or dietary shift? *Environ. Pollut.* **2013**, *172*, 216–222. [CrossRef] [PubMed]
49. Li Yung Lung, J.Y.S.; Tank, S.E.; Spence, C.; Yang, D.; Bonsal, B.; McClelland, J.W.; Holmes, R.M. Seasonal and geographic variation in dissolved carbon biogeochemistry of rivers draining to the Canadian Arctic Ocean and Hudson Bay. *J. Geophy. Res. Biogeosci.* **2018**, *123*, 3371–3386. [CrossRef]
50. Delaigue, L.; Thomas, H.; Mucci, A. Spatial variations in CO_2 fluxes in the Saguenay Fjord (Quebec, Canada) and results of a water mixing model. *Biogeosciences* **2020**, *17*, 547–566. [CrossRef]
51. Loux, N.T. An assessment of mercury-species-dependent binding with natural organic carbon. *Chem. Speciat. Bioavailab.* **1998**, *10*, 127–136. [CrossRef]
52. Haitzer, M.; Aiken, G.R.; Ryan, J.N. Binding of mercury(II) to dissolved organic matter: The role of the mercury-to-DOC concentration ratio. *Environ. Sci. Technol.* **2002**, *36*, 3564–3570. [CrossRef]
53. Brigham, M.; Wentz, D.; Aiken, G.; Krabbenhoft, D. Mercury Cycling in Stream Ecosystems. 1. Water Column Chemistry and Transport. *Environ. Sci. Technol.* **2009**, *43*, 2720–2725. [CrossRef]
54. Benoit, H.P.; Gagné, J.A.; Savenkoff, C.; Ouellet, P.; Bourrassa, M.-N. *State of the Ocean Report for the Gulf of St Lawrence Integrated Management (GOSLIM) Area*; Canadian Manuscript Report and of Fisheries Aquatic Sciences No. 2986; Fisheries and Ocean Canada: Moncton, NB, Canada, 2012; p. 73.

55. Cossa, D.; Heimbürger, L.-E.; Pérez, F.F.; García-Ibáñez, M.I.; Sonke, J.E.; Planquette, H.; Lherminier, P.; Boutorh, J.; Cheize, M.; Menzel Barraqueta, J.L.; et al. Mercury distribution and transport in the North Atlantic Ocean along the Geotraces-GA01 transect. *Biogeosciences* **2018**, *15*, 2309–2323. [CrossRef]
56. Cossa, D.; Heimbürger, L.-E.; Sonke, J.E.; Planquette, H.; Lherminier, P.; García-Ibáñez, M.I.; Pérez, F.F.; Sarthou, G. Sources, recycling and transfer of mercury in the Labrador Sea (Geotraces-Geovide cruise). *Mar. Chem.* **2017**, *198*, 64–69. [CrossRef]
57. Bourduas Crouhen, V.; Siron, R.; Blondlot, A. *État des Lieux de Pêche et D'aquaculture au Québec en lien avec les Changements Climatiques*; Ouranos: Montréal, QC, Canada, 2017; p. 84. Available online: www.ouranos.ca/publication-scientifique/Etat_peches_aquaculture2017.pdf (accessed on 15 August 2020).
58. Guillou, E.; Cyr, C.; Laplante, J.-F.; Bourque, F.; Toupoint, N.; Tremblay, R. Commercial Performance of Blue Mussel (*Mytilus edulis*, L.) Stocks at a Microgeographical Scale. *J. Mar. Sci. Eng.* **2020**, *8*, 382. [CrossRef]
59. De Ladurantaye, R.; Desjardins, C.; Nadeau, R.; Vignault, Y.; Larue, J.-F. *Les Contaminants dans le Saint-Laurent: Bilan des Connaissances Dans le Système Aquatique Marin. Ministère des Pêches et des Océans*; Technique report; Gouvernement du Canada, Symposium sur le Saint-Laurent; Congrès de l'Association des biologistes du Québec: St. Lawrence, QC, Canada, N 220702; November 1989; p. 21.
60. Pellerin, J.; Amiard, J.-C. Comparison of bioaccumulation of metals and induction of metallothioneine in two marine bivalves (*Mytilus edulis* and *Mya arenaria*). *Comp. Biochem. Phys. Part C Toxicol. Pharmacol.* **2009**, *150*, 186–195. [CrossRef]
61. Garron, C.; Gagné, F.; Ernst, W.; Julien, G.; Bernier, M.; Caldwell, C. Mercury contamination of marine sediment and blue mussel (*Mytilus edulis*) in the vicinity of a mercury cell chlor-alkali plant in Dalhousie, New-Brunswick, Canada. *Water Qual. Res. J. Can.* **2005**, *40*, 1–15. [CrossRef]
62. Walker, T.R. Mercury concentrations in marine sediments near a former mercury cell chlor-alkali plant in eastern Canada. *Mar. Pollut. Bull.* **2016**, *107*, 398–401. [CrossRef] [PubMed]
63. Cossa, D. Chemical contaminants in the St. Lawrence Estuary and Saguenay Fjord. In *Oceanography of a Large-Scale Estuarine System: The St. Lawrence*; El-Sabh, M.I., Silverberg, N., Eds.; Springer: New York, NY, USA, 1990; pp. 240–268.
64. Cossa, D.; Knoery, J.; Boye, M.; Marusczak, N.; Thomas, B.; Courau, P.; Sprovieri, F. Oceanic mercury concentrations on both sides of the Strait of Gibraltar decreased between 1989 and 2012. *Anthropocene* **2019**, *29*, 100230. [CrossRef]
65. Amos, H.M.; Jacob, D.J.; Streets, D.G.; Sunderland, E.M. Legacy impacts of all-time anthropogenic emissions on the global mercury cycle: Global impacts of legacy mercury. *Glob. Biogeochem. Cycles* **2013**, *27*, 410–421. [CrossRef]
66. Gustin, M.S.; Bank, M.S.; Bishop, K.; Bowman, K.; Branfireun, B.; Chételat, J.; Eckley, C.S.; Hammerschmidt, C.R.; Lamborg, C.; Lyman, S.; et al. Mercury biogeochemical cycling: A synthesis of recent scientific advances. *Sci. Total Environ.* **2020**, *737*, 139619. [CrossRef]

applied sciences

MDPI

Article

Long-Term Trends in Regional Wet Mercury Deposition and Lacustrine Mercury Concentrations in Four Lakes in Voyageurs National Park

Mark E. Brigham [1,*], David D. VanderMeulen [2], Collin A. Eagles-Smith [3], David P. Krabbenhoft [4], Ryan P. Maki [5] and John F. DeWild [4]

1 U.S. Geological Survey, 2280 Woodale Drive, Mounds View, MN 55112, USA
2 National Park Service, 2800 Lake Shore Dr. E., Suite D, Ashland, WI 54806, USA; david_vandermeulen@nps.gov
3 U.S. Geological Survey, 3200 SW Jefferson Way, Corvallis, OR 97330, USA; ceagles-smith@usgs.gov
4 U.S. Geological Survey, 8505 Research Way, Middleton, WI 53562, USA; dpkrabbe@usgs.gov (D.P.K.); jfdewild@usgs.gov (J.F.D.)
5 National Park Service, Voyageurs National Park, 360 Highway 11 East, International Falls, MN 56649, USA; ryan_maki@nps.gov
* Correspondence: mbrigham@usgs.gov

Featured Application: Long-term monitoring of mercury in precipitation and lacustrine ecosystems provides insights into ecosystem responses as a result of mercury and sulfate emissions reductions.

Citation: Brigham, M.E.; VanderMeulen, D.D.; Eagles-Smith, C.A.; Krabbenhoft, D.P.; Maki, R.P.; DeWild, J.F. Long-Term Trends in Regional Wet Mercury Deposition and Lacustrine Mercury Concentrations in Four Lakes in Voyageurs National Park. *Appl. Sci.* **2021**, *11*, 1879. https://doi.org/10.3390/app11041879

Academic Editor: Stéphane Guédron

Received: 31 October 2020
Accepted: 7 February 2021
Published: 20 February 2021

Abstract: Although anthropogenic mercury (Hg) releases to the environment have been substantially lowered in the United States and Canada since 1990, concerns remain for contamination in fish from remote lakes and rivers where atmospheric deposition is the predominant source of mercury. How have aquatic ecosystems responded? We report on one of the longest known multimedia data sets for mercury in atmospheric deposition: aqueous total mercury (THg_{aq}), methylmercury ($MeHg_{aq}$), and sulfate from epilimnetic lake-water samples from four lakes in Voyageurs National Park (VNP) in northern Minnesota; and total mercury (THg) in aquatic biota from the same lakes from 2001–2018. Wet Hg deposition at two regional Mercury Deposition Network sites (Fernberg and Marcell, Minnesota) decreased by an average of 22 percent from 1998–2018; much of the decreases occurred prior to 2009, with relatively flat trends since 2009. In the four VNP lakes, epilimnetic $MeHg_{aq}$ concentrations declined by an average of 44 percent and THg_{aq} by an average of 27 percent. For the three lakes with long-term biomonitoring, temporal patterns in biotic THg concentrations were similar to patterns in $MeHg_{aq}$ concentrations; however, biotic THg concentrations declined significantly in only one lake. Epilimnetic $MeHg_{aq}$ may be responding both to a decline in atmospheric Hg deposition as well as a decline in sulfate deposition, which is an important driver of mercury methylation in the environment. Results from this case study suggest that regional- to continental-scale decreases in both mercury and sulfate emissions have benefitted aquatic resources, even in the face of global increases in mercury emissions.

Keywords: mercury; methylmercury; lakes; wet deposition

1. Introduction

Human activities have considerably increased the amount of mercury in the atmosphere and deposited into aquatic ecosystems [1]. Elevated mercury levels in fish have resulted in widespread fish-consumption advisories across the United States (U.S.), North America, and globally to protect human health [2], and also pose ecotoxicological risk to piscivorous wildlife [3]. High fish-mercury levels can occur anywhere, including remote, relatively pristine ecosystems where the predominant source of mercury is atmospheric deposition.

Although deposited primarily as inorganic mercury, within aquatic ecosystems some fraction of the total mercury load is converted to methylmercury [3]. Mercury in fish tissues is predominantly in the methylmercury form [3]; thus, factors that control methylmercury production in aquatic ecosystems are of interest. Two such factors, relevant to the current paper, are the availability of inorganic mercury and sulfate. Increased mercury loading to aquatic ecosystems results in a proportional increase in methylmercury production and accumulation by aquatic biota [4]. Furthermore, because sulfate reducing bacteria are important as methylators of mercury [5,6], elevated sulfate deposition also can result in increased methylmercury production in aquatic ecosystems [5,7]; it is likely, therefore, that methylmercury concentrations in aquatic ecosystems (including fish) are not only greater than background, preindustrial levels, but greater than one might expect with solely a proportional response to mercury inputs.

With an increased understanding of the global extent of atmospherically driven mercury contamination in aquatic ecosystems, the U.S. and Canada have undertaken significant efforts to reduce anthropogenic mercury emissions. Lake-coring evidence indicates that mercury inputs to remote, mid-continental lakes peaked in the 1960s–1970s and declined subsequently [8]. More recently, emissions reporting has shown substantial declines in U.S. mercury emissions of approximately 87 percent from 1990–2017 [9], with the largest decreases occurring in the mid 1990's following regulations limiting emissions from hospital and municipal waste incinerators and chlor-alkali facilities; similarly large declines in mercury emissions occurred in Canada over the same time frame [10] (Table 1).

Table 1. Trends in emissions of mercury in the U.S. [9] and in the Canadian provinces of Manitoba and Ontario [10], 1990–2017. [U.S. data were reported in short tons; Canadian data were reported in kilograms [kg]; all data have been converted to Mg yr^{-1}; percentage change is from 1990 to 2017.].

Year	1990	2000	2005	2008	2011	2014	2017	Percentage Change
			Air emissions of mercury, total for all sectors, Mg yr^{-1}					
United States	223	–	95	55	51	47	30	−86.6%
Canada	34.3	9.5	7.9	6.6	3.8	3.5	3.0	−91.2%
Total US and Canada	257	–	103	62	55	51	33	−87.2%

Perhaps equally beneficial, from a methylmercury production and bioaccumulation perspective, efforts to control acid precipitation during the 1970s and 1980s led to large decreases in emission and deposition of sulfur oxides across the U.S. since 1970 [11]. Consistent with this continental-scale trend, sulfate deposition within the Voyageurs National Park region has declined substantially. Data from the National Atmospheric Deposition Program's (NADP) site MN16 (Marcell, Minnesota) show annual wet sulfate deposition declining from about 10 kg ha^{-1} in 1980, to about 6 kg ha^{-1} in 2000, to about 2.4 kg ha^{-1} in 2018 (Figure S1). Drevnick et al. [12] have previously attributed declining fish-mercury levels in lakes at Isle Royale National Park (Lake Superior, USA) to declines in sulfate loading.

Although North American mercury emissions have declined sharply (Table 1 and [13]), and emissions have declined in several other regions as well [13], industrial development and associated increases in mercury emissions—particularly in Asia—have led to a net global increase in anthropogenic mercury emissions of 1.8 percent per year [14]. The relative importance of North American versus global emissions to local deposition in northern Minnesota is unclear. A recent modeling study found that North American emission reductions have a more pronounced effect on wet mercury deposition reductions in the eastern U.S., where there was a greater concentration of sources, than in comparatively remote northern Minnesota [15], but that effort did not focus on ecosystem responses.

A critical question arises: how have lacustrine ecosystems in midcontinental North America responded to regional and North American mercury emissions versus global emissions? This question is particularly important because the Minamata Convention on Mercury [16], a legally-binding international treaty to reduce mercury use and releases to the environment, is being implemented by 128 signatory nations. Thus, assessing ecosystem responses to contemporary mercury reductions may be informative in assessing treaty implementation.

A few other studies have attempted to answer this question. There is mixed evidence for ecosystems responses to changing emissions trajectories, including some literature showing declining fish-mercury levels in response to regional emission reductions [17] and other studies showing long-term (1970s–2000s) declines in fish-mercury concentrations followed by a leveling off or increase in concentrations in recent years [18,19]. Mercury in the feathers of bald eagle (*Haliaeetus leucocephalus*) nestlings have shown similar temporal trends within the region [20].

This paper addresses the question: how have aquatic ecosystems responded? Here, we report on one of the longest known paired data sets for mercury in atmospheric deposition; mercury, methylmercury, and key other chemical and physical parameters in several lacustrine systems; and total mercury in aquatic biota from the same lacustrine systems. This paper reports an updated analysis of trends in mercury deposition at two northern Minnesota NADP Mercury Deposition Network (MDN) sites, as well as trends in methylmercury and total mercury concentrations in epilimnetic lake water and in biota from four lakes in a minimally disturbed area within Voyageurs National Park, also in northern Minnesota. Through several earlier studies [21–24], and an ongoing collaboration between the U.S. Geological Survey (USGS) and the National Park Service (NPS), these lakes have been sampled for mercury since 2000 or 2001 (depending on the lake), making these among the longest running data sets that pair total mercury and methylmercury in lake water with biotic mercury. Given the paucity of long-term data sets on aqueous mercury in undisturbed ecosystems, the sensitivity of circumneutral, low-ionic-strength aquatic ecosystems to the effects of mercury and sulfate deposition [7,25], and widespread consumption advisories in this region, the Voyageurs National Park data set provides a useful case study to monitor ecosystem responses to changes in atmospheric inputs. Trends in regional wet deposition, epilimnetic mercury, and mercury in age-1 yellow perch (*Perca flavescens*) were previously reported through 2012 [24]. This paper examines trends through 2018. In the current paper, the longer time period necessitated a change in trend analysis to account for nonlinearity in the data; also, given a change in organisms collected for biomonitoring, we applied a statistical relationship between mercury in dragonfly larvae and in age-1 yellow perch in order to extend the perch record.

2. Materials and Methods

2.1. Study Area

The study area and methods have been previously described [24]. In brief, the four study lakes lie within Voyageurs National Park, a park covering 883 km^2 in northeastern Minnesota (Figure 1; Table 2). The four lakes are drainage lakes, three of them fed by a small stream discharging from an upstream lake; the fourth (Ryan) fed only by a small headwaters stream. Although the upstream lakes may exert some influence on chemistry of the study lakes, given the small size of the inflowing streams, and relatively long water renewal times (0.6–1.2 year), the lake chemistry likely is governed more by catchment and deposition to the lake surface than by inflows from upstream lakes. Land cover within the park is largely boreal forest, with thin soils, and outcrops of Precambrian bedrock common throughout the park [26]. The lakes have been sampled for methylmercury and total mercury in lake water since 2000 (Shoepack Lake) or 2001 (Brown, Peary, and Ryan lakes) as part of earlier studies [21–24].

No monitoring of mercury deposition occurs within the park's borders; the two nearest NADP/MDN sites (http://nadp.slh.wisc.edu/mdn/, accessed on 1 February 2021) are used to characterize wet deposition of mercury: sites MN16 (Marcell, Minnesota), approximately 120 km south-southwest of Voyageurs National Park and MN18 (Fernberg, Minnesota), approximately 120 km southeast of the park. Due to the lack of large urban or industrial centers, relatively flat topography (hence lack of orographic effects), and variable wind direction in the region, data from these two monitoring sites are expected to be representative of mercury and sulfate deposition in the region. We examined trends in wet Hg deposition from 1998–2018; trends in aqueous methymercury ($MeHg_{aq}$) and total mercury (THg_{aq}) from 2000 (Shoepack Lake) or 2001 (Brown, Ryan, and Peary lakes) through 2018; and trends in biotic THg as described below.

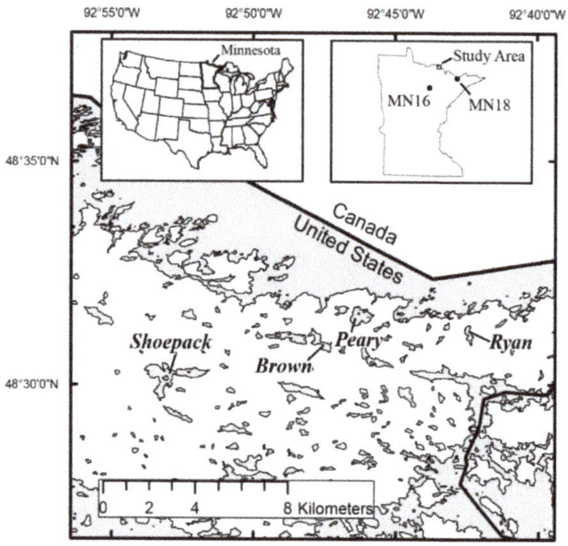

Figure 1. Location of four study lakes within Voyageurs National Park, and nearby National Atmospheric Deposition Program precipitation-monitoring sites MN16 and MN18.

Table 2. Selected lake characteristics [22,26]; and selected water-column measurements from the approximate centroid of each lake [24]. [W.A., watershed area; % Wetlands, wetlands as a percentage of watershed area; Renewal time = mean hydraulic residence time, in years; ANC, acid-neutralizing capacity; $\mu eq\ L^{-1}$, microequivalents per liter; TOC, total organic carbon; $mg\ L^{-1}$, milligrams per liter].

Lake Name	Lake Area (ha)	W.A. (ha)	% Wetlands	Lake Volume (10^6 m^3)	Maximum Depth (m)	Mean Depth (m)	pH	Renewal Time (yr)	ANC ($\mu eq\ L^{-1}$)	TOC ($mg\ L^{-1}$)	Secchi Transparency Depth (m)
Brown	30.8	158.9	9.9	1.3	8.2	4.2	7.0	0.8	129	9.4	2.7
Peary	45.3	734.6	19.8	1.2	4.6	2.6	7.2	0.6	195	10.1	2.4
Ryan	14.2	86.9	6.4	0.3	3.7	2.1	6.9	1.2	130	12.7	2.8
Shoepack	123.8	1735	25.8	3.6	7.3	2.9	6.5	0.7	114	15.5	1.3

2.2. Wet Deposition

Detailed methods are described by Prestbo et al. [27]. Briefly, weekly composited precipitation samples are collected and analyzed using modifications of U.S. Environmental Protection Agency (U.S.EPA) Methods 1669 [28] and 1631 [29]. Precipitation samples were analyzed for total mercury concentration (Hg_{precip}) at Frontier Global Sciences in Seattle, Washington. Data included herein met the NADP's acceptance criteria, [27,30] although some further screening of apparent outlier Hg_{precip} values was done as described below.

Inspection of weekly MDN data revealed a limited number of outliers in the weekly data, some of which were suspected as extreme outliers, i.e. unusually high or low, and which may impart an artefact into the calculations of annual wet Hg deposition rate and precipitation-weighted mean concentrations. We used the following approach to identify extreme outliers prior to wet deposition calculations:

Weekly data for the study sites were downloaded from the NADP/MDN website. Weekly Hg_{precip} and precipitation volumes were analyzed using a multiple-regression approach developed previously [31], using the REG procedure in SAS software (v. 9.4. SAS Institute, Carey, NC, USA). Hg_{precip} and precipitation depth data were log-transformed to remove skewness and heteroscedasticity of residuals. A regression model was developed to account for the exogenous effects of weekly precipitation depth; season (Fourier terms); and time (T, in years) (Equation (1)).

$$\log[Hg_{precip}] = \beta_0 + \beta_1 \log(Precip) + \beta_2 \sin(2\pi T) + \beta_3 \cos(2\pi T) + \beta_4 T + \varepsilon \quad (1)$$

where Hg_{precip} is mercury concentration in weekly precipitation samples in ng L^{-1}; Precip is the precipitation depth in mm; sin and cos are the sine and cosine functions; T is time in decimal years. Additional Fourier terms (sine and cosine of $4\pi T$) were significant at many MDN sites across North America [31], but were not significant for the current study sites, thus were not included in the analysis herein. The residual, ε, is the unexplained variation in $\log[Hg_{precip}]$, accounting for exogenous effects of precipitation amount, seasonality, and time.

Inspection of the data indicated the presence of outliers. Therefore, two passes of the regression analysis were performed. After the first pass, Hg_{precip} in weekly samples was set to missing if deemed an outlier according to the following rules:

Rule 1: absolute value of studentized residual from multiple-regression analysis (Equation (1)) greater than 4, regardless of precipitation amount. This removed extreme outliers for weekly samples with less than 10 mm of precipitation; low-volume events tend to be noisier. From this analysis, four outlier results were removed for MN16 and three for MN18.

Rule 2: for samples with precipitation volume >10 mm, values were set to missing if the absolute value of studentized residual from multiple-regression analysis (Equation (1)) was greater than 3. Using this rule, one Hg_{precip} result was set to missing for MN16 and two for MN18.

Mercury concentrations for the screened samples were deemed extreme outliers for the precipitation amount of the event and were likely an artefact rather than a realistic representation of Hg_{precip} in precipitation. One outlier in particular had an abnormally high Hg_{precip} for a relatively large precipitation event and occurred late in the study period (site MN18; 5 July 2017; reported Hg_{precip} = 89.59 ng L^{-1}; Precip = 27.686 mm). The resultant calculated weekly Hg deposition was 2.48 μg m^{-2}, which is about 25 percent of the annual wet deposition reported by NADP/MDN at this site and would have influenced the trend analysis were it not removed. A table of removed outliers is provided (Supplementary Materials, Table S1).

The above-described screening process retained weekly observations (with precipitation volume retained), but outlier values of Hg_{precip} were set to missing. After removal of outliers, regression analysis was applied (again using Equation (1)), and missing $\log[Hg_{precip}]$ values were predicted by the regression. Predicted Hg_{precip} was calculated by exponentiating regression-predicted $\log[Hg_{precip}]$ values, and multiplying by the mean of the exponentiated residuals to correct for back-transformation bias (Duan smearing estimator, as described in Helsel et al., pp. 256–257 [32]). Weekly Hg_{precip} values, then, consisted of MDN-reported values in nearly all cases, but regression-predicted values where weekly Hg_{precip} was missing (including observations where outliers were screened as described above).

Annual wet Hg deposition was then calculated as the sum of weekly deposition rates (concentration times precipitation depth); and precipitation-weighted mean Hg_{precip} were calculated as annual deposition rate divided by annual precipitation depth. This methodology differs from NADP/MDN network methodology. NADP/MDN effectively estimates missing Hg_{precip} as the precipitation-weighted annual mean concentration, whereas our methodology estimates missing Hg_{precip} using regression-based predictions based on the precipitation volume and season associated with each weekly sample. For the purposes of this paper, regression-based estimates were preferred because, given the dependence of Hg_{precip} concentrations on precipitation depth, precipitation-weighted mean concentrations may underestimate concentrations for small events and overestimate concentrations for large events (which are more important in determining annual loads).

Owing to modest nonlinearity in the time trends, trends in the annual wet Hg deposition rate and in precipitation-weighted annual mean Hg concentration were determined by locally weighted regression, after dealing with extreme outlier Hg concentrations as described above. Locally weighted regression used the LOESS procedure in SAS software, allowing the procedure to select the smoothing parameter by Akaike information criterion algorithm.

2.3. Lake Water

Lake water sampling and analytical methods have been described previously [24,33]. Unfiltered, epilimnetic lake water was sampled two to three times (typically three) per year, between May and September, during most years of the study period. Field crews generally sampled the upper ca. 5 cm of water out of a canoe heading upwind, in the approximate center of the lake, using trace-metal clean sampling protocols described in [21]. Water was collected in a pre-cleaned Teflon® FEP (fluorinated ethylene propylene) bottle, preserved by addition of HCl (to a final normality of ca. 0.02 *N*), and shipped to the USGS Mercury Research Laboratory (MRL) in Middleton, Wisconsin, for analysis. Detailed descriptions of all the analytical procedures used by the USGS MRL are available at the following website: https://wi.water.usgs.gov/mercury-lab/research (accessed on 1 February 2021) and the descriptions are summarized below. Total Hg determinations in lake water (THg_{aq}) were determined by U.S.EPA Method 1631 [29]. The USGS MRL typically achieves a daily detection limit (DDL) for total mercury analytical runs of about 0.06 pM and the precision, measured as the relative percent difference (RPD) between analytical duplicates averages 10 percent. Methylmercury ($MeHg_{aq}$) samples were analyzed by U.S.EPA Method 1630 ([34,35] with the added advancement (starting in 2006) of including isotope dilution by adding a small known amount (about 30 picograms) of isotopically labeled methylmercury ($Me^{200}Hg$) to each sample, which allows for more accurate measures of sample recovery rates.

Water samples for other constituents were also collected from the same lake locations with a 2 m long, 3.2 cm inner diameter PVC tube that integrates a 2 L sample from the upper 2 m of the water column. Samples collected once per summer, typically in July, were analyzed for several major ions including sulfate by White Water Associates, Inc. (Amasa, Michigan) (2006–2013) and CT Laboratories (Baraboo, Wisconsin) (2014–2018). Sample processing, handling, and quality assurance and quality control procedures are described in Elias et al. [36].

For the 2001–2012 time period [24], trends in $MeHg_{aq}$ conformed reasonably to a linear-regression model for three of the four lakes. However, inspection of data for the longer time (through 2018) revealed nonlinearity at some sites. Therefore, we used locally weighted regression using the LOESS procedure in SAS, again allowing the procedure to select the smoothing parameter by the default Akaike Information Criterion algorithm within the LOESS procedure.

To calculate the percent change in concentrations from 2001–2018, we used LOESS-predicted concentrations for arbitrary dates approximately at the beginning and end of the period of data collection (1 July 2001 and 1 July 2018). No slope parameter (and hence, no *p*-value for significance) is calculated in LOESS. Relevance of the trends can be assessed by examining the magnitude of change viewed along with the variability of the data; and by examining whether the 95-percent confidence interval for the LOESS smooth at the end of the time period includes, or does not include, the LOESS-predicted value for the start of the time period.

Lake water chemistry and lake level data from 2000–2007 are available from the USGS's National Water Information System web retrieval (https://doi.org/10.5066/F7P55KJN, accessed on 1 February 2021), for the following USGS site identification numbers: Brown

Lake (483059092474501); Peary Lake (483129092462001); Ryan Lake (483109092422601); and Shoepack Lake (482951092531601). Starting in 2006, sample collection and data archiving was led by the NPS, and data are available for retrieval from the National Water Quality Monitoring Council's (NWQMC) water-quality data portal (https://www.waterqualitydata.us/portal/, accessed on 1 February 2021), using the organization ID of 11NPSWRD_WQX, and the following site identification numbers: Brown Lake (11NPSWRD_WQX-VOYA_12); Peary Lake (11NPSWRD_WQX-VOYA_14); Ryan Lake (11NPSWRD_WQX-VOYA_17); and Shoepack Lake (11NPSWRD_WQX-VOYA_05). Data for mercury in yellow perch, and dragonfly larvae from 2008–2012 are also available at the NWQMC's water-quality portal for the same sites. Dragonfly larvae (Odonata, Anisoptera) data from 2014–2018 are available within a USGS data release [37].

2.4. Lake Levels

In 2006, the NPS established reference points on the shore of each lake. After establishing these, lake levels were determined relative to an arbitrary datum at each lake's reference point. Water-level anomaly was calculated as the difference between water level on a given sampling date and the initial water level relative to local datum.

2.5. Mercury in Lake Biota

Details of sampling for age-1 yellow perch and dragonfly larvae have been described elsewhere [24,38,39]. In brief, both yellow perch and dragonfly larvae were sampled annually from each lake during spring. Because total mercury concentrations (THg) can vary among families [38], we normalized THg in dragonfly larvae to those of a single family (Aeshnidae) following Eagles-Smith et al. [38]. This ensures a consistent unit of dragonfly larvae THg for each site year. Fish sampling occurred from 2000 to 2012, whereas dragonfly larvae were sampled from 2009 to 2018. To facilitate temporal comparisons across the entire study period we first examined the relation in THg concentrations between paired samples of yellow perch and Aeshnid-equivalent dragonfly larvae from 14 lakes in national parks in the western Great Lakes region because previous findings have shown them to be correlated [40], using linear regression of the geometric mean THg concentrations of each taxa where they were collected together. This analysis indicated that dragonfly larvae THg concentrations well correlated with those in yellow perch (*See Results*); therefore, we used the linear regression equation to estimate yellow perch THg concentrations for years where yellow perch were not sampled.

As with aqueous MeHg concentrations, examination of the temporal trends in yellow perch THg revealed nonlinearity in the three lakes with a complete temporal data set (Brown Lake, Ryan Lake, and Parry Lake). Therefore, we similarly used the LOESS procedure as described above to estimate the change in biotic Hg concentrations over time, though the lower data density necessitated a higher degree of smoothing than with the higher resolution $MeHg_{aq}$ sampling.

3. Results

3.1. Wet Deposition Trends

Rates of both annual wet Hg deposition and precipitation-weighted mean Hg concentrations at both MN16 and MN18 decreased over the time period 1998–2018 (Figure 2; Table 3); much of the decreases occurred prior to 2009, with relatively flat trends since 2009. Of interest from an ecosystem perspective is the change in wet-depositional loading. For reasons previously noted, removal of outliers, then use of regression-predicted weekly Hg_{precip} values to calculate weekly and annual wet Hg deposition rates, was preferable to using NADP/MDN-reported annual deposition rates. Expected values, from locally weighted regression analysis, of annual wet deposition rates decreased by 20 and 24 percent, respectively for MN16 and MN18, over the 1998–2018 period. These percentage declines are smaller than those reported for 1998–2012 [24]; the change in magnitude may be due to a different analytical approach and effect of removing influential outliers, as well as a flattening of the trends starting around 2009. The flatter trends since ~2009 may reflect a leveling-off of emissions-reductions in the U.S. and Canada over the last decade; i.e., large reductions in mercury emissions occur early in this time period, with more modest reductions in recent years (Table 1).

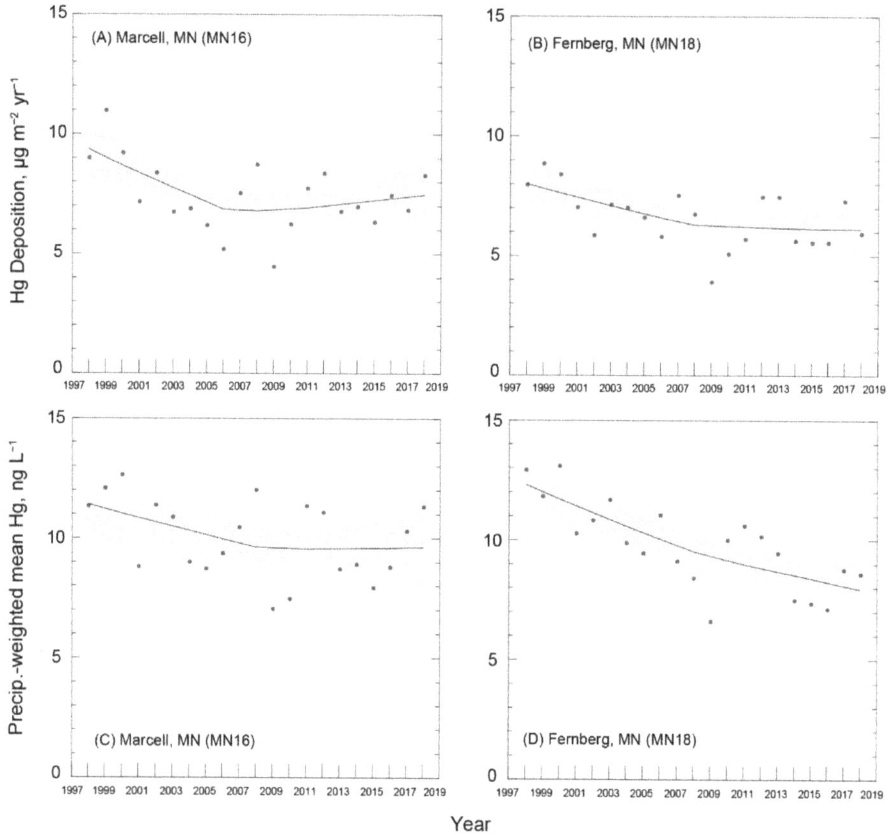

Figure 2. Trend plots of annual wet mercury (Hg) deposition rates for MDN sites MN16 (**A**) and MN18 (**B**); and precipitation-weighted annual mean Hg concentrations for MDN sites MN16 (**C**) and MN18 (**D**) (Trend line is locally weighted regression; shaded region is 95-percent confidence interval. LOESS smoothing parameters are as follows: (**A**) 0.833; (**B**) = 1.0; (**C**) 1.0; and (**D**) 1.0.

Because precipitation amount is used in the calculation of wet Hg deposition rate, a trend in precipitation could drive a trend in Hg deposition. Linear regression of precipitation depth versus time shows no significant trend for MN16 (p = 0.80) and a weak positive trend at MN18 (p = 0.10) (see Figure S2: precipitation volume trend plots). The sites display considerable interannual variability in total precipitation depth (ranges: 554–908 and 508–832 mm yr^{-1} for MN16 and MN18, respectively), driving interannual variability in wet Hg deposition. The weak, positive trend in precipitation amount at MN18 likely drove the relatively larger decline in precipitation-weighted mean concentrations at that site, compared to MN16, as larger precipitation events tend to have lower Hg_{precip}.

Given the lack of a significant trend in precipitation volume, we conclude that the observed overall trend of declining wet Hg deposition rate is likely driven by reductions of mercury emissions in North America and not trends in precipitation. In addition, the observed declines in mercury deposition are synchronous with known declines in North

American mercury emissions since 1990 [13,41], although global emissions have been comparatively constant [13]. A recent modeling study indicated that in northern Minnesota, emission reductions in North America are roughly equally important in comparison to emission reductions in the rest of the world in determining wet Hg deposition trends [15].

Table 3. Trend analysis results for wet mercury (Hg) deposition rate and precipitation-weighted mean concentrations for two National Atmospheric Deposition/Mercury Deposition Network sites in northern Minnesota. [Hg deposition µg m^{-2}, annual Hg deposition rate in micrograms per square meter; Precip.-weighted [Hg], ng L^{-1}, annual precipitation-weighted mean Hg concentration in nanograms per liter; EV_{1998} and EV_{2018} are expected values for 1998 and 2018, respectively, from locally weighted regression of the variable of interest versus time in years. % change, percentage change in expected values from 1998–2018. Lower and upper 95 percent confidence limits shown in brackets.].

	Hg Deposition, µg m^2			Precip.-Weighted [Hg], ng L^{-1}		
MDN/NTN Site	**EV_{1998}**	**EV_{2018}**	**% Change**	**EV_{1998}**	**EV_{2018}**	**% Change**
Marcell (MN16)	9.37	7.47	−20%	11.41	9.63	−16%
	[8.05, 10.70]	[6.14, 8.80]		[9.88, 12.95]	[8.10, 11.17]	
Fernberg (MN18)	8.00	6.12	−24%	12.31	7.93	−36%
	[6.99, 9.01]	[5.11, 7.12]		[11.12, 13.50]	[6.74, 9.13]	
	Mean percent change:		−22%			−26%

3.2. Trends in Epilimnetic Methylmercury, Total Mercury, and Sulfate Concentrations

Both methylmercury and total mercury concentrations declined in epilimnetic lake water over the 2001–18 period, although the declines at some lakes were small, relative to variability. For Brown Lake, a high methylmercury outlier previously identified [24] was omitted from trend analysis. The overall decline in $MeHg_{aq}$ for Brown Lake is modest (32%; Table 4), but noteworthy for two reasons. First, the previously reported trend for Brown Lake [24] was positive, but weak; and second, concentrations have declined sharply since peaking around 2010, about the end of the time frame for the previous trend analysis. Peary and Ryan Lakes had the largest declines in $MeHg_{aq}$ (Table 4), with much of the change occurring in the first few years of record. The comparatively large declines in these two lakes is consistent with earlier findings [24]. Shoepack Lake also exhibited a modest decline in $MeHg_{aq}$. The magnitude of the declines in $MeHg_{aq}$ for Brown and Shoepack Lakes is modest, in relation to the variability in concentrations at these two lakes; in addition, the 95-percent confidence intervals for the start and end of the period of study overlap for these two lakes suggesting that the decline is not significant (Figure 3).

Table 4. Epilimnetic lake water trends in methylmercury ($MeHg_{aq}$), total mercury (THg_{aq}), and sulfate, from locally weighted regression analysis, for the period 2001–2018. (EV_{2001} and EV_{2018} are expected values for concentrations of $MeHg_{aq}$ and THg_{aq}, in ng L^{-1}; and sulfate, in mg L^{-1}, from locally weighted regression analysis, for 1 July 2001, and 1 July 2018, respectively. % change, percentage change in expected values from 2001–2018. Lower and upper 95-percent confidence limits shown in brackets).

	$MeHg_{aq}$			THg_{aq}			Sulfate		
	EV_{2001}	**EV_{2018}**	**% Change**	**EV_{2001}**	**EV_{2018}**	**% Change**	**EV_{2001}**	**EV_{2018}**	**% Change**
Brown	0.0854	0.0582	−32%	1.94	1.53	−21%	1.31	0.88	−33%
	[0.037, 0.134]	[0.011, 0.106]		[1.57, 2.32]	[1.16, 1.90]		[0.95, 1.66]	[0.58, 1.17]	
Peary	0.1706	0.0468	−73%	1.58	1.11	−29%	1.24	0.71	−43%
	[0.140, 0.201]	[0.018, 0.076]		[1.26, 1.89]	[0.81, 1.41]		[0.33, 2.15]	[−0.05, 1.46]	
Ryan	0.2380	0.1243	−48%	3.35	1.78	−47%	2.79	1.66	−41%
	[0.194, 0.281]	[0.086, 0.162]		[2.53, 4.17]	[1.08, 2.50]		[2.31, 3.26]	[1.26, 2.05]	
Shoepack	0.2592	0.1941	−25%	2.73	2.44	−11%	1.63	0.61	−62%
	[0.182, 0.337]	[0.108, 0.281]		[2.28, 3.18]	[1.85, 3.02]		[0.66, 2.59]	[−0.17, 1.39]	
		mean = −44%			mean = −27%			mean = −45%	

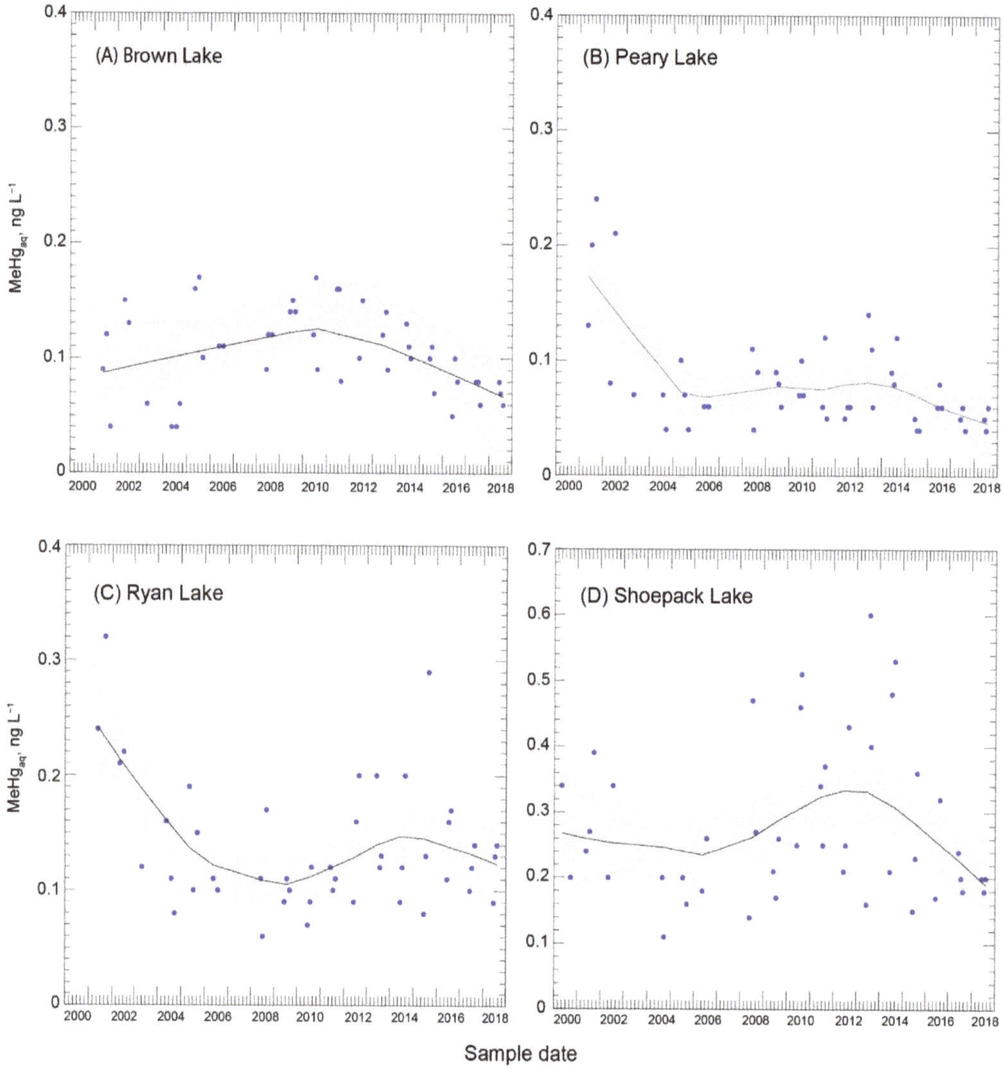

Figure 3. Time series plots for methylmercury ($MeHg_{aq}$) in epilimnetic lake water. Smooth lines are locally weighted regression lines; smoothing parameters, selected as described in the methods section: (**A**) Brown Lake, smoothing parameter = 0.713; (**B**) Peary Lake, smoothing parameter = 0.424; (**C**) Ryan Lake, smoothing parameter = 0.489; and (**D**) Shoepack Lake, smoothing parameter = 0.633. Gray shading indicates the 95-percent confidence interval. A high outlier for Brown Lake (reported concentration 0.51 ng L^{-1} for 30 August 2012) is omitted from the plot and regression.

Aqueous total mercury declined modestly in Brown, Peary, and Shoepack Lakes, with an overlap of the 95-percent confidence intervals for the start and end period of the study (again, indicating perhaps a lack of statistical significance) (Figure 4). The 47% decline in THg_{aq} in Ryan Lake appears to be a significant decline with clear separation of confidence intervals in the beginning versus end of the study period.

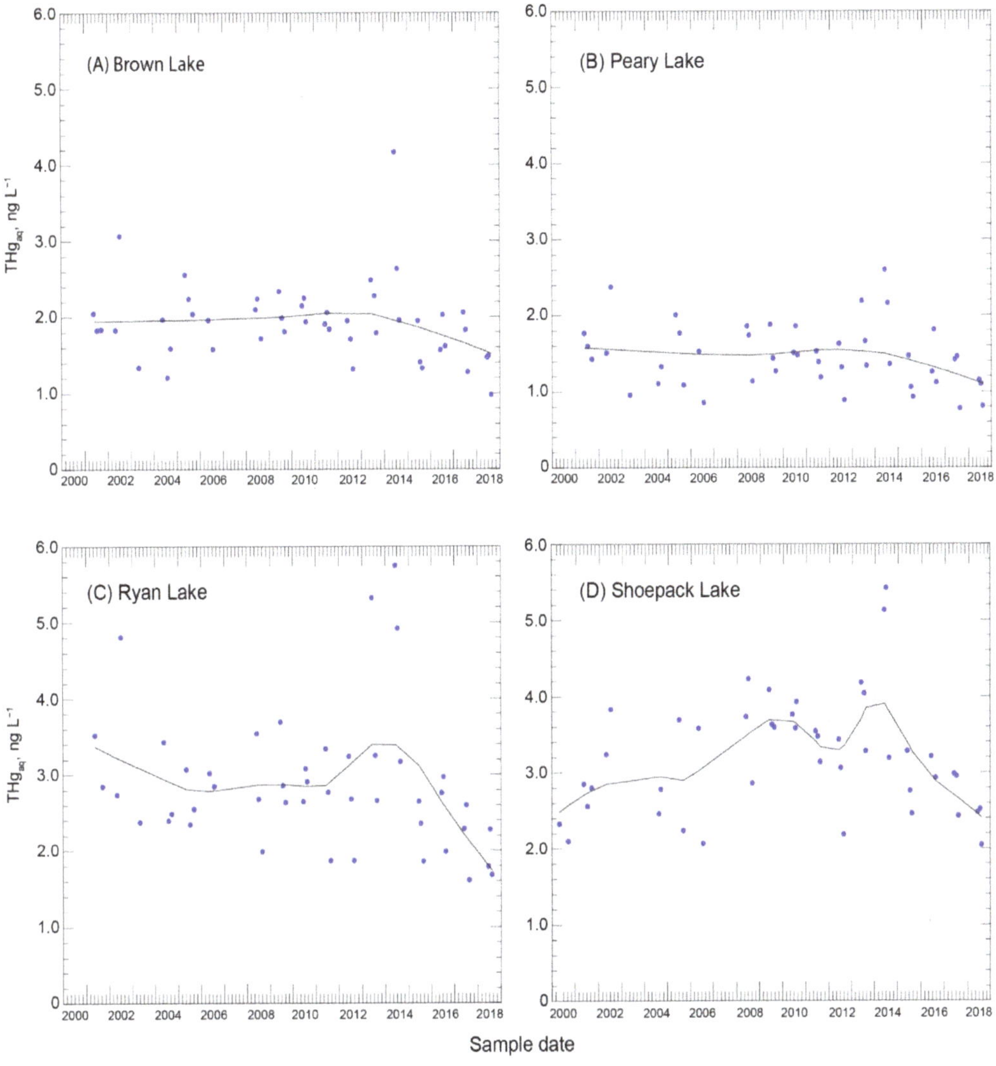

Figure 4. Time series plots for total mercury (THg$_{aq}$) in epilimnetic lake water. Smooth lines are locally weighted regression lines; smoothing parameters, selected as described in the methods section: (**A**) Brown Lake, smoothing parameter = 0.713; (**B**) Peary Lake, smoothing parameter = 0.641; (**C**) Ryan Lake, smoothing parameter = 0.424; (**D**) Shoepack Lake, smoothing parameter = 0.3. Gray shading indicates the 95-percent confidence interval.

Epilimnetic sulfate concentrations decreased in each lake over the study period (Table 4; Figure 5). The trends were nonlinear, revealing midtime series peak concentrations around 2007 for Brown, Ryan, and Shoepack Lakes, and somewhat later (~2010) for Peary Lake. The modest trends reported here (mean decrease of 45%) follow much larger decreases for these lakes from the 1980s to 2000, as reported by Kallemeyn et al. [26], based on a 1980s lake survey reported by Payne [42].

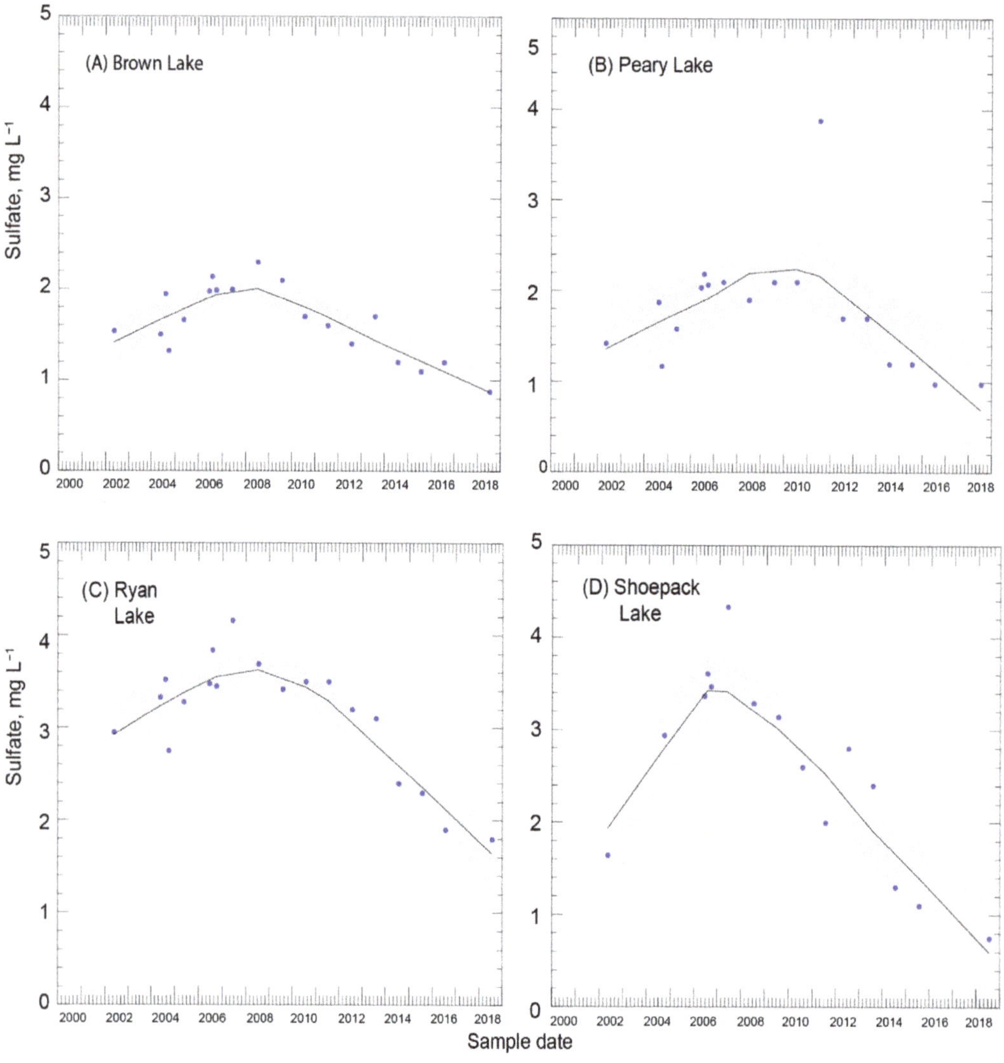

Figure 5. Time series plots for sulfate in epilimnetic lake water. Smooth lines are locally weighted regression lines; smoothing parameters, selected as described in the methods section: (**A**) Brown Lake, smoothing parameter = 0.711; (**B**) Peary Lake, smoothing parameter = 0.75; (**C**) Ryan Lake, smoothing parameter = 0.711; and (**D**) Shoepack Lake, smoothing parameter = 0.70. Gray shading indicates the 95-percent confidence interval.

3.3. Lake Level Fluctuations

$MeHg_{aq}$ correlates modestly (R^2 = 0.36) with lake-level anomaly for Brown Lake (Figure 6). However, none of the other lakes in this study showed significant correlations between $MeHg_{aq}$ and lake-level anomaly (not shown).

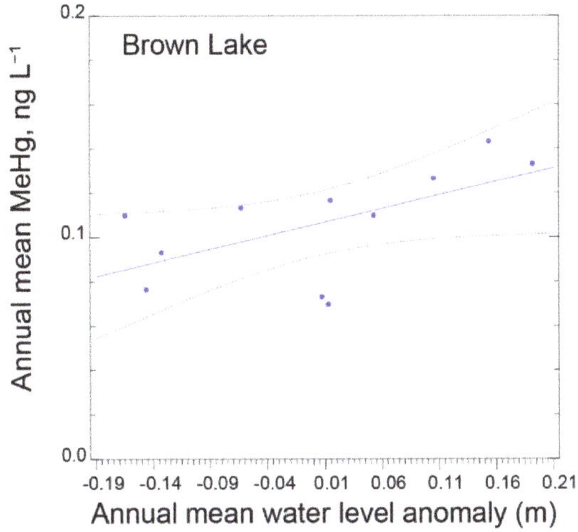

Figure 6. Plot of annual mean methylmercury ($MeHg_{aq}$) concentration for epilimnetic lake water, versus annual mean water-level anomaly relative to a local datum, for Brown Lake. (R^2 = 0.36, p = 0.04; 95-percent confidence interval shown as dashed line.).

3.4. *Mercury Trends in Lake Biota*

Total mercury concentrations in age-1 yellow perch were well correlated with Aeshnid-equivalent dragonfly larvae THg concentrations ($R^2 = 0.66$, $p < 0.0001$, N = 40; Figure 7), facilitating converting dragonfly THg concentrations to those of yellow perch for years when fish were not sampled. Yellow perch data for all four lakes were collected through 2012 as summarized previously [24]. After yellow perch collections ceased, only Brown, Peary, and Ryan Lakes were sampled for dragonfly larvae, so only those three lakes are considered here.

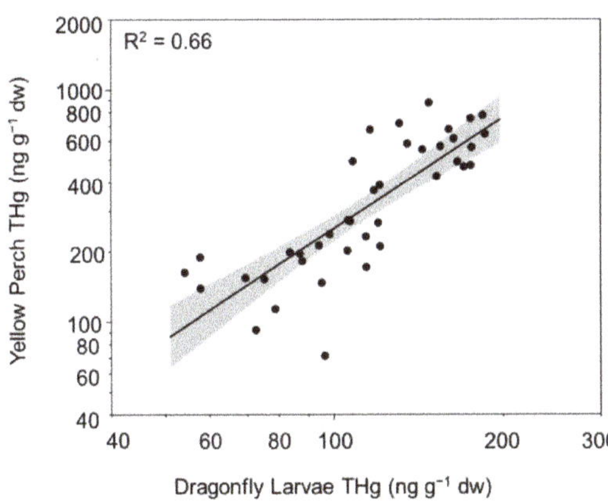

Figure 7. Total mercury (THg) concentrations in age-1 yellow perch in relation to dragonfly larvae (Aeshnid equivalent) THg concentrations. Shaded area shows the 95-percent confidence interval. [Regression equation: $\ln(THg_{YPE}) = -1.756 + 1.582 \times \ln(THg_{AE})$, $r^2 = 0.66$, $p < 0.0001$, N = 40, where ln is the natural logarithm; THg_{YPE}, total mercury in age-1 yellow perch equivalent; THgAE, total mercury in Aeshnid equivalent.].

For the three lakes with long-term biomonitoring, temporal patterns in biotic THg concentrations were similar to patterns in $MeHg_{aq}$ concentrations (Figures 3 and 8); however, biotic THg concentrations declined in only Peary Lake. Expected values for yellow perch THg for Brown Lake increased by 4.6% between 2000 and 2018, but the 95-percent confidence intervals overlapped between those years indicating that the difference is not significant; similar to $MeHg_{aq}$, there was a substantial 54% increase in THg concentrations in yellow perch between 2000 and 2010, followed by a 46% decrease between 2010 and 2018. As with $MeHg_{aq}$, Peary Lake had the greatest decline in biotic THg (45%), which was primarily driven by the 31% decrease between 2000 and 2010. Ryan Lake showed an initial decline of biotic THg until about 2010, followed by an increase; yellow perch THg concentrations in Ryan Lake increased by 5% between 2000 and 2018, similar in magnitude to Brown Lake, and not apparently significant. However, in contrast with Brown Lake, at Ryan Lake there was a substantial decline (38%) between 2000 and 2010, followed by a 69% increase between 2010 and 2018—a considerably sharper increase than $MeHg_{aq}$ concentrations during the same time period.

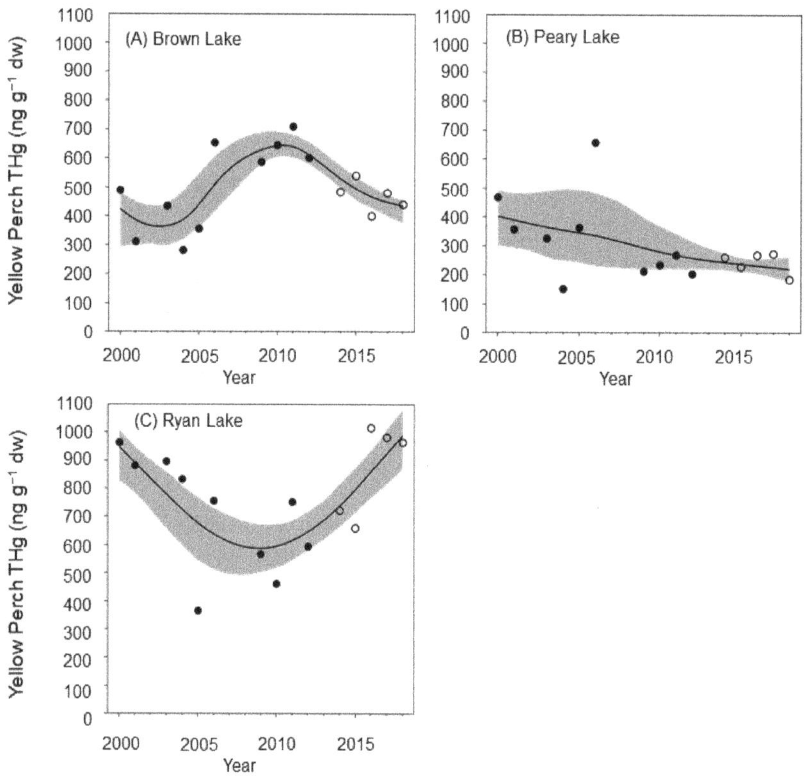

Figure 8. Total mercury (THg) in age-1 yellow perch versus time in (**A**) Brown Lake, (**B**) Peary Lake, and (**C**) Ryan Lake. Solid symbols represent geometric mean concentrations from yellow perch, open symbols indicate that yellow perch concentrations are derived from dragonfly larvae using the linear regression shown in Figure 7. Shaded area shows the 95-percent confidence interval. [Smooth lines are locally weighted regression. Smoothing parameters: (**A**) Brown Lake = 0.75; (**B**) Peary Lake = 0.83; and (**C**) Ryan Lake = 0.76. Gray shading indicates the 95-percent confidence interval.

4. Discussion

Wet Hg deposition at two regional MDN sites (Fernberg and Marcell, Minnesota) declined by an average of 22 percent from 1998–2018, with much of the decline occurring prior to 2010. In the four lakes, epilimnetic $MeHg_{aq}$ concentrations declined by an average of 44 percent and THg_{aq} by an average of 27 percent. Although the magnitude of trend in some lakes was small, it is noteworthy that for all the lakes both $MeHg_{aq}$ and THg_{aq} show declines for the 2001–2018 time period, including the latter part of that period when wet Hg deposition rates leveled off, suggesting a lag related to watershed inputs. Epilimnetic $MeHg_{aq}$ may be responding both to a decline in atmospheric Hg deposition as well as a decline in sulfate deposition, which is an important driver of mercury methylation in the environment. The long-term reduction in epilimnetic sulfate concentrations in the lakes also reflects declines in sulfate deposition, as has been observed elsewhere [43]. This observation is a good example of the importance of collecting data on other known key factors (for example, sulfate) that control mercury cycling in the environment when the goal is to accurately attribute the drivers of change.

Environmental mercury data sets that include long-term monitoring of multiple media (atmosphere, surface water, and biota) in a relatively small area are exceedingly rare. As such, the opportunities to assess baselines and trends in mercury levels in aquatic ecosystems, especially in the lead-up to expected globally driven emissions changes from the Minamata Treaty [16], are likewise uncommon.

Previously, it was hypothesized that inflowing water from a lake upstream from Brown Lake (Oslo Lake), which yielded relatively high concentrations of $MeHg_{aq}$ in a 2001–2002 survey of small lakes in Voyageurs National Park [21], could be responsible for the increase in $MeHg_{aq}$ in Brown Lake during the 2001–2012 time period. The apparent trend reversal, i.e., the decline in $MeHg_{aq}$ in Brown Lake that occurred starting around 2010, coupled with the correlation between lake level and $MeHg_{aq}$, supports the hypothesis.

Higher observed $MeHg_{aq}$ concentrations coincident in time with higher lake levels is consistent with the generally held conceptual understanding from the mercury literature that wetter conditions and cyclical inundation and draining of low-lying areas (such as wetlands) leads to increased $MeHg_{aq}$ production within a lake's watershed, irrespective of loading from upstream lakes. However, the remaining lakes in this study showed no correlation between $MeHg_{aq}$ and lake level. This null finding is in contrast to the coherence of water level and MeHg in water and fish observed for lakes in northern Wisconsin [44,45]. However, it should be noted that the ecological setting in northern Wisconsin is quite different from the Voyageurs National Park region, especially in regard to hydrology. The Northern Highlands of northern Wisconsin are characterized by high permeability due to deep outwash sands and gravel that yield poorly integrated surface drainages. Our study area, in contrast, is more of a classical boreal system with shallow soils overlying bedrock and highly integrated flow systems. As such, the lack of concurrence between findings of Watras et al. [44,45] and our study is not surprising.

Trends in fish-tissue THg concentrations moderately tracked $MeHg_{aq}$ or THg_{aq} for Brown and Peary Lakes but not Ryan Lake. Whereas $MeHg_{aq}$ concentrations often correspond to biological mercury uptake and accumulation in many water bodies, there can be substantial variability in the efficiency of transfer into and through food webs due to the context dependence of site-specific bio-geochemical and ecological drivers. Also, whereas $MeHg_{aq}$ or THg_{aq} is an instantaneous measure of conditions, biological tissues integrate exposure over much longer time periods, including the Odonates that are generally several years old. This disparity can complicate interpretations of the effectiveness of decreasing mercury emissions and deposition. In addition, for boreal-like settings, the connectivity to terrestrial soils and their legacy accumulation of decades of mercury deposition is well understood; however, the internal time lags of how long this large pool of mercury will continue to yield meaningful amounts of mercury to downstream aquatic ecosystems remains unknown. This finding does not imply that declining mercury emissions and deposition (and subsequent $MeHg_{aq}$ production) offer limited benefits for mercury risk reduction. Instead, it emphasizes the need to interpret long-term environmental mercury data sets in the context of a complex set of pathways and processes that control mercury cycling in the environment, and the need for multimedia (deposition, water, and biota) and multiconstituent data (beyond just mercury and methylmercury measurements) for effective trend analysis for mercury.

Results from this case study suggest that regional- to continental-scale decreases in both mercury and sulfate emissions have benefitted aquatic resources, even in the face of global increases in mercury emissions. The reductions in atmospheric pollutant loading may be of considerable benefit to human and ecosystem health, considering that mercury-based fish-consumption advisories are in place for all lakes of Voyageurs National Park and many lakes in the region, and northern pike (*Esox lucius*) mercury levels in park lakes have exceeded thresholds for detrimental effects to fish reproduction [46,47].

A number of MDN sites across North America had substantial declines in wet mercury deposition from the late 1990s through early 2000s, followed by a leveling-off and in some cases increase in Hg deposition starting around 2010 [31]. The two northern Minnesota sites appear to fit this broader geographic pattern. Trend analysis by locally weighted regression showed a relatively sharp decline in the period from 1998 to about 2010, followed by a leveling-off of deposition rates. Although the MDN has a data review and quality assurance program in place, the data-screening procedure employed herein identified a small number of extreme outlier mercury concentrations that appeared unreasonable. Because extreme outliers for an individual sample can bias annual wet-deposition calculations, our screening procedure (or similar ones) warrants further consideration.

The larger declines in epilimnetic $MeHg_{aq}$, compared to epilimnetic THg_{aq} is likely driven by both decline in wet Hg deposition (and thus in-lake THg_{aq}), as well as declines in sulfate deposition. As noted previously [24], in response to emission controls related to the Clean Air Act, sulfate deposition has declined dramatically in northern Minnesota, as well as more broadly across North America [11,48]. Other research has shown that adding sulfate to wetlands greatly increases methylmercury production [7], whereas decreased sulfate loading results in decreased methylmercury production [49].

Owing to long-term atmospheric deposition of anthropogenic mercury and sulfate, lakes in Voyageurs National Park, and regionally, surely have elevated methylmercury levels in both water and biota, relative to pre-industrial conditions. It is encouraging, however, that declines in anthropogenic mercury and sulfur emissions have translated to declines in wet mercury and sulfate deposition, which in turn appear to have resulted in declines in methylmercury contamination in lake ecosystems within the park. The relatively large $MeHg_{aq}$ declines, in comparison to declines in THg_{aq}, are consistent with the notion that $MeHg_{aq}$ levels are influenced by both anthropogenic mercury as well as anthropogenic sulfate deposition. It is worth noting that the response of lake ecosystems to decreased mercury inputs is expected to include both a rapid component owing to direct deposition to the lake surface and a slow component driven as previously deposited mercury slowly re-equilibrates from wetland and upland soils [50].

Lastly, as emphasized previously, there are relatively few published long-term, multimedia data sets that include atmospheric mercury-deposition monitoring coupled with methylmercury and total mercury in lake water and mercury in lake biota. This is particularly important for undisturbed settings where methylmercury production and bioaccumulation are largely governed by natural processing of atmospheric pollutant loads. Watras et al. [44] reported trends for aqueous total mercury and methylmercury and biotic mercury for two lakes in northern Wisconsin (Little Rock Lake, 1988–2017 and Trout Bog, 1999–2017) that are relatively close (ca. 275 km southeast of VNP), yet the two studies yielded trend analyses that are notably different. This variability in temporal trends is consistent with the overarching influence that within-lake and watershed bio-geochemistry can have on mercury transport and methylmercury production, potentially decoupling them from trends in mercury loading. This highlights the importance of considering the context of each ecosystem and supports the notion that recovery from many decades of sustained mercury emissions is unlikely to be a linear process. Data sets like the one used in this study, while rare, will serve as critically important baselines for executing effectiveness evaluations associated with the post-Minamata-Treaty implementation. Although more extensive networks have been proposed to monitor ecosystem responses to controls on anthropogenic mercury emissions [51], in the absence of such programs leading up to the global change expected from the Minamata Treaty, researchers might better coordinate small-scale, long-term research efforts so that broader-scale assessments can be made.

Supplementary Materials: The following are available online at https://www.mdpi.com/2076-3417/11/4/1879/s1, Table S1: Outlier samples identified by regression analysis of log[Hg_{precip}] versus logPrecip, seasonal terms, and time. Figure S1: Annual wet sulfate deposition for the National Atmospheric Deposition Network site at Marcell, Minnesota. Figure S2: Precipitation volume trend plots for National Atmospheric Deposition Network/Mercury Deposition Network sites Marcell, Minnesota (MN16) and Fernberg, Minnesota (MN18).

Author Contributions: M.E.B. analyzed precipitation and lake-water chemistry data; C.A.E.-S. analyzed the biotic data. J.F.D. contributed to water chemistry methods and quality assurance of the water chemistry data. All authors contributed to preparation of the manuscript. All authors have read and agreed to the published version of the manuscript.

Funding: Preparation of this manuscript was funded by the National Park Service (NPS). National Atmospheric Deposition Network/Mercury Deposition Network sites were supported by the U.S. Forest Service, U.S. Environmental Protection Agency, and the Minnesota Pollution Control Agency (MPCA). Funding earlier research that contributed data to the current paper is detailed in [24]. The University of Wisconsin at LaCrosse's collection of some of the biotic data was funded by the Great Lakes Restoration Initiative. Partial financial support for lake water and biotic chemistry data was provided by the MPCA and the U.S. Geological Survey (USGS). Sampling of yellow perch was funded by the NPS Great Lakes Inventory and Monitoring Network under Task Agreement J2105080012 of the Great Lakes-Northern Forest Cooperative Ecosystem Studies Unit and by the Great Lakes Restoration Initiative, Environmental Protection Agency Project Number 222, under Task Agreement J2105100001 of the Great Lakes-Northern Forest Cooperative Ecosystem Studies Unit—both under Cooperative Agreement H6000082000 between the NPS and the University of Minnesota. The USGS/NPS Water Quality Partnership supported lake-water data collection in 2001–03. The USGS Toxic Substances Hydrology program supported the analysis of samples and staff time at the USGS Mercury Research Laboratory to process the samples, archive the data, and assist in the data interpretation.

Institutional Review Board Statement: Previous fish research incorporated into this study was conducted in accordance with Animal Care and Use Procedure (ACUP 306.02) published by Cornell University and adopted by researchers from the University of Wisconsin-La Crosse and the U.S. National Park Service.

Informed Consent Statement: Not applicable.

Data Availability Statement: Data used for this study are available, as detailed in the Methods section.

Acknowledgments: Since 2008, lake sampling was conducted by National Park Service staff, particularly Jaime LeDuc, with Richard Damstra and Alex Egan assisting with data management. In addition, since 2008, mercury analyses of lake-water samples have been provided by the U.S. Geological Survey's Mercury Research Laboratory (MRL); Jacob Ogorek (USGS/MRL) is acknowledged for laboratory support. Assistance with the sampling and analyses of fish and dragonfly larvae was provided by University of Wisconsin at La Crosse students including Sean Bailey, Matthew Brantner, Adam Hyer, John Kalas, Kevin Miller, Kristen Thoreson, Jeffrey Ziegeweid; and faculty Roger Haro, Mark Sandheinrich, and James Wiener (retired). We also thank the Dragonfly Mercury Project team (Colleen Flanagan Pritz, Sarah Nelson, James Willacker, Amanda Klemmer, Meghan Hess, Colleen Emery) and USGS Contaminant Ecology Lab (Branden Johnson, John Pierce) for laboratory support and data management. Jennifer Murphy (USGS) offered helpful advice on statistics. We thank Austin Baldwin (U.S. Geological Survey) and three anonymous reviewers for helpful reviews of earlier versions of this manuscript. Any use of trade, firm, or product names is for descriptive purposes only and does not imply endorsement by the U.S. Government.

Conflicts of Interest: The authors declare no conflict of interest.

References

1. Swain, E.B.; Engstrom, D.R.; Brigham, M.E.; Henning, T.A.; Brezonik, P.L. Increasing rates of atmospheric mercury deposition in midcontinental North America. *Science* **1992**, *257*, 784–787. [CrossRef] [PubMed]
2. U.S. Environmental Protection Agency. 2010 Biennial National Listing of Fish Advisories. 2011. Available online: https://archive.epa.gov/ow/ost/web/html/technical_factsheet_2010.html (accessed on 1 February 2020).
3. Wiener, J.G.; Krabbenhoft, D.P.; Heinz, G.H.; Scheuhammer, A.M. Ecotoxicology of mercury, Chapter 16. In *Handbook of Ecotoxicology*, 2nd ed.; Hoffman, D.J., Rattner, B.A., Burton, G.A., Jr., Cairns, J., Jr., Eds.; CRC Press: Boca Raton, FL, USA, 2003; pp. 409–463.
4. Orihel, D.M.; Paterson, M.J.; Blanchfield, P.J.; Bodaly, R.A.; Hintelmann, H. Experimental evidence of a linear relationship between inorganic mercury loading and methylmercury accumulation by aquatic biota. *Environ. Sci. Technol.* **2007**, *41*, 4952–4958. [CrossRef]
5. Gilmour, C.C.; Henry, E.A.; Mitchell, R. Sulfate stimulation of mercury methylation in freshwater sediments. *Environ. Sci. Technol.* **1992**, *26*, 2281–2287. [CrossRef]
6. Compeau, C.G.; Bartha, R. Sulfate-reducing bacteria—principal methylators of mercury in anoxic estuarine sediment. *Appl. Environ. Microbiol.* **1985**, *50*, 498–502. [CrossRef] [PubMed]
7. Jeremiason, J.D.; Engstrom, D.R.; Swain, E.B.; Nater, E.A.; Johnson, B.M.; Almendinger, J.E.; Monson, B.A.; Kolka, R.K. Sulfate addition increases methylmercury production in an experimental wetland. *Environ. Sci. Technol.* **2006**, *40*, 3800–3806. [CrossRef] [PubMed]
8. Engstrom, D.R.; Swain, E.B. Recent declines in atmospheric mercury deposition in the Upper Midwest. *Environ. Sci. Technol.* **1997**, *31*, 960–967. [CrossRef]
9. U.S. Environmental Protection Agency. 2017 National Emissions Inventory Technical Support Document. 2020; p. 486. Available online: https://www.epa.gov/sites/production/files/2020-04/documents/nei2017_tsd_full_30apr2020.pdf (accessed on 1 October 2020).
10. Government of Canada. Air Pollutants Emissions Inventory Online Search. 2020. Available online: https://pollution-waste.canada.ca/air-emission-inventory (accessed on 10 March 2020).
11. Nopmongcol, U.; Beardsley, R.; Kumar, N.; Knipping, E.; Yarwood, G. Changes in United States deposition of nitrogen and sulfur compounds over five decades from 1970 to 2020. *Atmos. Environ.* **2019**, *209*, 144–151. [CrossRef]
12. Drevnick, P.E.; Canfield, D.E.; Gorski, P.R.; Shinneman, A.L.C.; Engstrom, D.R.; Muir, D.C.G.; Smith, G.R.; Garrison, P.J.; Cleckner, L.B.; Hurley, J.P.; et al. Deposition and cycling of sulfur controls mercury accumulation in Isle Royale fish. *Environ. Sci. Technol.* **2007**, *41*, 7266–7272. [CrossRef] [PubMed]
13. Zhang, Y.; Jacob, D.J.; Horowitz, H.M.; Chen, L.; Amos, H.M.; Krabbenhoft, D.P.; Slemr, F.; Louis, V.L.S.; Sunderland, E.M. Observed decrease in atmospheric mercury explained by global decline in anthropogenic emissions. *Proc. Natl. Acad. Sci. USA* **2016**, *113*, 526–531. [CrossRef]
14. Streets, D.G.; Horowitz, H.M.; Lu, Z.; Levin, L.; Thackray, C.P.; Sunderland, E.M. Global and regional trends in mercury emissions and concentrations, 2010–2015. *Atmos. Environ.* **2019**, *201*, 417–427. [CrossRef]
15. Giang, A.; Song, S.; Muntean, M.; Janssens-Maenhout, G.; Harvey, A.; Berg, E.; Selin, N.E. Understanding factors influencing the detection of mercury policies in modelled Laurentian Great Lakes wet deposition. *Environ. Sci. Process. Impacts* **2018**, *20*, 1373–1389. [CrossRef] [PubMed]
16. United Nations Environment Programme. Minamata Convention on Mercury. 2020. Available online: https://www.mercuryconvention.org/ (accessed on 1 February 2021).
17. Hutcheson, M.S.; Smith, C.M.; Rose, J.; Batdorf, C.; Pancorbo, O.; West, C.R.; Strube, J.; Francis, C. Temporal and spatial trends in freshwater fish tissue mercury concentrations associated with mercury emissions reductions. *Environ. Sci. Technol.* **2014**, *48*, 2193–2202. [CrossRef]
18. Monson, B.; Staples, D.; Bhavsar, S.; Holsen, T.; Schrank, C.; Moses, S.; McGoldrick, D.; Backus, S.; Williams, K. Spatiotemporal trends of mercury in walleye and largemouth bass from the Laurentian Great Lakes Region. *Ecotoxicology* **2011**, *20*, 1555–1567. [CrossRef] [PubMed]
19. Gandhi, N.; Tang, R.W.K.; Bhavsar, S.P.; Arhonditsis, G.B. Fish Mercury Levels Appear to Be Increasing Lately: A Report from 40 Years of Monitoring in the Province of Ontario, Canada. *Environ. Sci. Technol.* **2014**, *48*, 5404–5414. [CrossRef]
20. Dykstra, C.R.; Route, W.T.; Williams, K.A.; Meyer, M.W.; Key, R.L. Trends and patterns of PCB, DDE, and mercury contamination in bald eagle nestlings in the upper Midwest. *J. Great Lakes Res.* **2019**, *45*, 252–262. [CrossRef]
21. Goldstein, R.M.; Brigham, M.E.; Stuewe, L.; Menheer, M.A. Mercury Data from Small Lakes in Voyageurs National Park, Northern Minnesota, 2000–02. 2003; U.S. Geological Survey; Open-File Report 03-480; p. 18. Available online: http://pubs.er.usgs.gov/publication/ofr2003480 (accessed on 1 February 2021). [CrossRef]
22. Wiener, J.G.; Knights, B.C.; Sandheinrich, M.B.; Jeremiason, J.D.; Brigham, M.E.; Engstrom, D.R.; Woodruff, L.G.; Cannon, W.F.; Balogh, S.J. Mercury in soils, lakes, and fish in Voyageurs National Park (Minnesota)—importance of atmospheric deposition and ecosystem factors. *Environ. Sci. Technol.* **2006**, *40*, 6261–6268. [CrossRef]
23. Woodruff, L.G.; Sandheinrich, M.; Brigham, M.E.; Cannon, W.F. Impact of wildfire on levels of mercury in forested watershed systems–Voyageurs National Park, Minnesota. 2009; U.S. Geological Survey; Scientific Investigations Report 2009-5151; p. 19. Available online: http://pubs.usgs.gov/sir/2009/5151/ (accessed on 1 February 2021).

24. Brigham, M.E.; Sandheinrich, M.B.; Gay, D.A.; Maki, R.P.; Krabbenhoft, D.P.; Wiener, J.G. Lacustrine responses to decreasing wet mercury deposition rates—results from a case study in northern Minnesota. *Environ. Sci. Technol.* **2014**, *48*, 6115–6123. [CrossRef]
25. Sonesten, L. Catchment Area Composition and Water Chemistry Heavily Affects Mercury Levels in Perch (*Perca Fluviatilis* L.) in Circumneutral Lakes. *Water Air Soil Pollut.* **2003**, *144*, 117–139. [CrossRef]
26. Kallemeyn, L.W.; Holmberg, K.L.; Perry, J.A.; Odde, B.Y. *Aquatic Synthesis for Voyageurs National Park*; Information and Technology Report 2003-0001; U.S. Geological Survey: International Falls, MN, USA, 2003; p. 96. Available online: http://www.cerc.usgs.gov/pubs/center/pdfdocs/ITR2003-0001.pdf (accessed on 1 February 2021).
27. Prestbo, E.M.; Gay, D.A. Wet deposition of mercury in the U.S. and Canada, 1996-2005: Results and analysis of the NADP mercury deposition network (MDN). *Atmos. Environ.* **2009**, *43*, 4223–4233. [CrossRef]
28. U.S. Environmental Protection Agency. *Method 1669—Sampling Ambient Water for Trace Metals at EPA Water Quality Criteria Levels*; U.S. Environmental Protection Agency: Washington, DC, USA, 1996; p. 35, EPA 821/R-96-011.
29. U.S. Environmental Protection Agency. *Method 1631, Revision E—Mercury in Water by Oxidation, Purge and trap, and Cold Vapor Atomic Fluorescence Spectrometry*; U.S. Environmental Protection Agency: Washington, DC, USA, 2002; p. 35, EPA-821-R-02-019.
30. Vermette, S.; Bloom, N.; Tokos, J.; Welker, M.; Verry, S.; Lindberg, S. *The Mercury Deposition Network (NADP/MDN): Transition Phase, February 1995 to February 1996*; National Atmospheric Deposition Program, Illinois State Water Survey: Champaign, IL, USA, 1996.
31. Weiss-Penzias, P.S.; Gay, D.A.; Brigham, M.E.; Parsons, M.T.; Gustin, M.S.; Schure, A.T. Trends in mercury wet deposition and mercury air concentrations across the U.S. and Canada. *Sci. Total Environ.* **2016**, *568*, 546–556. [CrossRef] [PubMed]
32. Helsel, D.R.; Hirsch, R.M.; Ryberg, K.R.; Archfield, S.A.; Gilroy, E.J. *Statistical Methods in Water Resources: U.S. Geological Survey Techniques and Methods, Book 4, Chapter A3*; U.S. Geological Survey: Reston, VA, USA, 2020; p. 458. [CrossRef]
33. VanderMeulen, D.D.; Lafrancois, B.M.; Edlund, M.B.; Hobbs, J.M.R.; Damstra, R. Pairing modern and paleolimnological approaches to evaluate the nutrient status of lakes in Upper Midwest National Parks. *J. Am. Water Resour. Assoc.* **2016**, *52*, 1401–1419. [CrossRef]
34. DeWild, F.J.; Olson, M.L.; Olund, S.D. *Determination of Methyl Mercury by Aqueous Phase Ethylation, Followed by Gas Chromatographic Separation with Cold Vapor Atomic Fluorescence Detection*; Open-File Report 01-445; U.S. Geological Survey: Denver, CO, USA, 2002; p. 14. Available online: http://pubs.usgs.gov/of/2001/ofr-01-445/ (accessed on 1 February 2021).
35. U.S. Environmental Protection Agency. *Method 1630—Methyl Mercury in Water by Distillation, Aqueous Ethylation, Purge and Trap, and Cold Vapor Atomic Fluorescence Spectrometry*; U.S. Environmental Protection Agency: Washington, DC, USA, 1998; p. 55, EPA-821-R-01-020.
36. Elias, J.E.; Axler, R.; Ruzycki, E.; VanderMeulen, D. *Water Quality Monitoring Protocol for Inland Lakes: Great Lakes Inventory and Monitoring Network, Version 1.1*; National Park Service: Fort Collins, CO, USA, 2015; p. 384. Available online: https://irma.nps.gov/DataStore/Reference/Profile/2224131 (accessed on 1 February 2021).
37. Eagles-Smith, C.A.; Nelson, S.J.; Flanagan-Pritz, C.M.; Willacker, J.J., Jr.; Klemmer, A. Total Mercury Concentrations in Dragonfly Larvae from U.S. National Parks (ver. 4.0, 20 October 2020). U.S. Geological Survey: Corvallis, OR, USA, 2018. [CrossRef]
38. Eagles-Smith, C.A.; Willacker, J.J.; Nelson, S.J.; Pritz, C.M.F.; Krabbenhoft, D.P.; Chen, C.Y.; Ackerman, J.T.; Grant, E.H.C.; Pilliod, D.S. A National-Scale Assessment of Mercury Bioaccumulation in United States National Parks Using Dragonfly Larvae As Biosentinels through a Citizen-Science Framework. *Environ. Sci. Technol.* **2020**, *54*, 8779–8790. [CrossRef] [PubMed]
39. Haro, R.J.; Bailey, S.W.; Northwick, R.M.; Rolfhus, K.R.; Sandheinrich, M.B.; Wiener, J.G. Burrowing Dragonfly Larvae as Biosentinels of Methylmercury in Freshwater Food Webs. *Environmental Environ. Sci. Technol.* **2013**, *47*, 8148–8156. [CrossRef]
40. Wiener, J.G.; Haro, R.J.; Rolfhus, K.R.; Sandheinrich, M.B.; Bailey, S.W.; Northwick, R.M.; Gostomski, T.J. *Bioaccumulation of Persistent Contaminants in Fish and Larval Dragonflies in Six National Park Units of the Western Great Lakes Region, 2008–2009*; National Park Service: Fort Collins, CO, USA, 2013; p. 55, Natural Resource Data Series NPS/GLKN/NRDS—2013/427. Available online: https://irma.nps.gov/DataStore/Reference/Profile/2192699 (accessed on 1 February 2021).
41. Schmeltz, D.; Evers, D.; Driscoll, C.; Artz, R.; Cohen, M.; Gay, D.; Haeuber, R.; Krabbenhoft, D.; Mason, R.; Morris, K.; et al. MercNet—A national monitoring network to assess responses to changing mercury emissions in the United States. *Ecotoxicology* **2011**, *20*, 1713–1725. [CrossRef] [PubMed]
42. Payne, G.A. *Water Quality of Lakes and Streams in Voyageurs National Park, Northern Minnesota, 1977–84*; Water-Resources Investigations Report 88-4016; U.S. Geological Survey: St. Paul, MN, USA, 1991; p. 95. Available online: http://pubs.usgs.gov/wri/1988/4016/report.pdf (accessed on 1 February 2021).
43. Driscoll, C.T.; Driscoll, K.M.; Fakhraie, H.; Civerolo, K. Long-term temporal trends and spatial patterns in the acid-base chemistry of lakes in the Adirondack region of New York in response to decreases in acidic deposition. *Atmos. Environ.* **2016**, *146*, 5–14. [CrossRef]
44. Watras, C.J.; Grande, D.; Latzka, A.W.; Tate, L.S. Mercury trends and cycling in northern Wisconsin related to atmospheric and hydrologic processes. *Can. J. Fish. Aquat. Sci.* **2019**, *76*, 831–846. [CrossRef]
45. Watras, C.J.; Teng, H.-Y.; Latzka, A.W.; Meyer, M.W.; Zhang, Z. Near-Decadal Oscillation of Water Levels and Mercury Bioaccumulation in the Laurentian Great Lakes Region. *Environ. Sci. Technol. Lett.* **2020**, *7*, 89–94. [CrossRef]
46. Sandheinrich, M.B.; Wiener, J.G. Methylmercury in freshwater fish—recent advances in assessing toxicity of environmentally relevant exposures. In *Environmental Contaminants in Biota—Interpreting Tissue Concentrations*, 2nd ed.; Beyer, W.N., Meador, J.P., Eds.; Taylor and Francis: Boca Raton, FL, USA, 2011; pp. 169–190.

47. Wiener, J.G.; Haro, R.J.; Rolfhus, K.R.; Sandheinrich, M.B.; Bailey, S.W.; Northwick, R.M.; Gostomski, T.J. *Bioaccumulative Contaminants in Aquatic Food Webs in Six National Park units of the Western Great Lakes Region—2008–2012*; Natural Resource Report NPS/GLKN/NRR—2016/1302; National Park Service: Fort Collins, CO, USA, 2016; p. 136. Available online: https://irma.nps.gov/DataStore/Reference/Profile/2233537 (accessed on 1 February 2021).
48. Lehmann, C.M.B.; Gay, D.A. Monitoring long-term trends of acidic wet deposition in US precipitation—results from the National Atmospheric Deposition Program. *Power Plant Chem.* **2011**, *13*, 378–385.
49. Coleman Wasik, J.K.; Mitchell, C.P.J.; Engstrom, D.R.; Swain, E.B.; Monson, B.A.; Balogh, S.J.; Jeremiason, J.D.; Branfireun, B.A.; Eggert, S.L.; Kolka, R.K.; et al. Methylmercury declines in a boreal peatland when experimental sulfate deposition decreases. *Environ. Sci. Technol.* **2012**, *46*, 6663–6671. [CrossRef] [PubMed]
50. Harris, R.C.; Rudd, J.W.M.; Amyot, M.; Babiarz, C.L.; Beaty, K.G.; Blanchfield, P.J.; Bodaly, R.A.; Branfireun, B.A.; Gilmour, C.C.; Graydon, J.A.; et al. Whole-ecosystem study shows rapid fish-mercury response to changes in mercury deposition. *Proc. Natl. Acad. Sci. USA* **2007**, *104*, 16586–16591. [CrossRef]
51. Harris, R.; Krabbenhoft, D.P.; Mason, R.; Murray, M.W.; Reash, R.; Saltman, T. (Eds.) *Ecosystem Responses to Mercury Contamination—Indicators of Change*; CRC Press: Boca Raton, FL, USA, 2007; p. 216.

applied sciences

MDPI

Article

Oxidative Stress Biomarkers in Erythrocytes of Captive Pre-Juvenile Loggerhead Turtles Following Acute Exposure to Methylmercury

Javier Hernández-Fernández [1,2,*], Ellie Anne López-Barrera [3], Leonardo Mariño-Ramírez [4], Pilar Rodríguez-Becerra [1] and Andrés Pinzón-Velasco [5]

1 Department of Natural and Environmental Sciences, Faculty of Science and Engineering, Genetics, Molecular Biology and Bioinformatic Research Group—GENBIMOL, Jorge Tadeo Lozano University, Cra. 4 No 22-61, Bogotá 110311, Colombia; mariad.rodriguezb@utadeo.edu.co
2 Department of Biology, Faculty of Sciences, Pontificia Universidad Javeriana, Calle 45, Cra. 7, Bogotá 110231, Colombia
3 Sergio Arboleda University, Institute of Environmental Studies and Services, IDEASA Research Group—IDEASA, Bogotá 111711, Colombia; ellie.lopez@usa.edu.co
4 NCBI, NLM, NIH Computational Biology Branch, Building 38A, Room 6S614M 8600 Rockville Pike, MSC 6075 Bethesda, MD 20894-6075, USA; marino@ncbi.nlm.nih.gov
5 Bioinformatics and Systems Biology Group, National University of Colombia, Calle 45 Cra. 30, Bogotá 111321, Colombia; ampinzonv@unal.edu.co
* Correspondence: javier.hernandez@utadeo.edu.co

Received: 21 April 2020; Accepted: 19 May 2020; Published: 22 May 2020

Abstract: This study describes the use of erythrocytes (RBCs) of loggerhead turtles as in vitro models for evaluating their toxicity to methylmercury. Blood samples of loggerhead turtles that were born in the Colombian Caribbean were used. The LC_{50} of RBCs to methylmercury was determined at 96 h using methylmercury concentrations of 0.5–100 mg L^{-1}. Next, the viability of the RBCs and the activity of the enzymes superoxide dismutase (SOD), glutathione S-transferase (GST), and lipid peroxidation by malondialdehyde (MDA) at 6 and 12 h of exposure to acute concentrations of 0, 1, and 5 mg L^{-1} were evaluated. The LC_{50} for loggerhead turtle RBCs was 8.32 mg L^{-1}. The cell viability bioassay of RBCs exposed for 12 h only showed 100% cell viability. Increasing in vitro MeHg concentrations caused a corresponding increase in MDA concentration as well as decreases in the activities of SOD and GST. The RBCs represent an excellent model for ecotoxicological studies and SOD, GST, and MDA are biomarkers of environmental pollution and oxidative stress in loggerhead turtles. This was the first study conducted on loggerhead turtle where the response of RBCs to MeHg-induced oxidative stress is evaluated.

Keywords: Cytotoxicity; erythrocytes; methylmercury; malondialdehyde; in vitro; superoxide dismutase

1. Introduction

The loggerhead turtle (*Caretta caretta*) is a marine reptile that is distributed throughout the world; however, its population is in decline and in danger of extinction due to anthropic and environmental causes. The loggerhead turtle is considered to be a sentinel species of environmental pollution due to its longevity and high rate of biomagnification of contaminants [1]. These conditions allow for loggerhead turtles to be used as monitoring units for environmental pollutants. Loggerhead turtles reside in the Mediterranean Sea, and the Atlantic, Indian, and Pacific oceans [2]. In Colombia, the loggerhead turtle is distributed in the departments of Atlántico, San Andrés Archipelago, Providencia and Santa Catalina, La Guajira, Magdalena, and Sucre. It has been reported that some juvenile individuals of this species use the Colombian Caribbean as foraging area, where they feed on algae, vascular plants, and invertebrates, such as crabs, mollusks, shrimps, starfish, and jellyfish [3]. It has been estimated that sea turtles can survive for very long periods (more than 50 years), time during which they store tremendously contaminating heavy metals in their tissues. Among them, mercury enters through the digestive tract and it is accumulated [4]. These sea turtles are sensitive to mercury storage due to their life history (prolonged anoxia, longevity, omnivore feeding, aquatic habitat) [4]. In Colombia, the loggerhead turtle is categorized as critically endangered according to the criteria of the IUCN. A 97% reduction of individuals was reported, from 200 nesting females that arrived per season to the Colombian Caribbean (in 1972) to six per year in the most recent assessment [5]. For this reason, it is important to preserve this specie in order to carry out studies that increase the survival, birth rates, reproductive success, and allow for collecting samples to monitor their health, and support future research [6].

Mercury produces lethal and sublethal effects in wildlife, causing different alterations (immunopathological, carcinogenic, teratogenic, neural, cardiovascular, renal, reproductive, and endocrine) [7–10]. To date, it is known that mercury, interacting with the sulfhydryl groups of cells, generates oxidative stress and alters calcium levels. However, the reaction that it experiences in its methylated state (MeHg), with hundreds of proteins and the changes that they cause, need to be investigated in greater depth in order to understand the adverse effects on embryonic and fetal development, and how it interferes with the normal functioning of the neurological, cardiovascular, and immune systems [11]. Mercury (Hg) has potential for biomagnification in the environment. Its high ecotoxicological risk is due to its biogeochemical behavior, allowing for it to associate with solids, suspended matter, and sediments [10]. In oceans and seas, mercury is present in its inorganic and organic forms, with the methylated form, being methylmercury, the most toxic one (CH_3-Hg^1+) (MeHg) [12]. MeHg is more toxic than inorganic mercury, and it prevails in the marine food chain. It typically comprises 75–99% of the total mercury (Hg-T) in organisms [13].

In the Colombian Caribbean, sediment Hg-T concentrations range between 20.6–552.7 μg L^{-1}, with averages around 180–300 μg L^{-1} and peaks of up to 1300 μg L^{-1} [14,15]. The threshold effect level of Hg-T is 130 μg L^{-1} [15]. It is estimated that the global release of mercury from natural sources, anthropogenic sources, and reemission from past activities is 6000–11,000 tons/year [11]. In particular, anthropogenic sources are believed to account for approximately 80% of the annual Hg emission to the environment [16,17].

Methylmercury induces the production of reactive oxygen species (ROS), generating oxidative stress and altering the antioxidant defense system of the cells when it joins the sulfhydryl and selenotiol groups [18,19]. It also affects the permeability of phospholipid membranes, by producing porosity, stiffness, and lipid peroxidation [19]. Organisms have an antioxidant metabolic machinery made up of the enzymes glutathione-s-transferase (GST), glutathione peroxidase (GPX), superoxide dismutase (SOD), glutathione reductase (GR), and others, which neutralize free radicals and other ROS and prevent cell damage, to neutralize ROS that is produced by oxidative stress induced by the action of MeHg [20]. In red blood cells, glutathione in its reduced and oxidized states, together with the coenzyme NADPH, help to restore the $Fe+^2$ state in hemoglobin [21].

Erythrocytes (RBCs), unlike other cell types, have a high affinity for methylmercury [22]. In addition, they express a large amount of enzymatic and non-enzymatic antioxidants allowing for the monitoring of the antioxidant response [23]. RBCs are nucleated in birds, reptiles, fish, and amphibians. These blood cells have been used in environmental studies stress [24–30], and to evaluate the response of xenobiotics in mammals [31–33]. However, hardly any studies have been published where environmental pollutants have been quantified in sea turtles [34,35]. To date, there are no reports of the use of RBCs in specific bioassays in which oxidative stress markers were evaluated for exposure to environmental pollutants in sea turtles.

The present study aims to evaluate the use of erythrocytes (RBCs) of captive loggerhead turtles as a predictive in vitro model of oxidative stress that is produced by acute exposure to different concentrations of MeHg. The Median Lethal Concentration (LC_{50}) and the viability of the RBCs were determined and, with this result, the concentrations of the bioassay were selected, evaluating SOD, GST, and Malonaldehyde (MDA) activity. These results provide a novel, non-invasive, cost-effective method for ecototoxicological studies on the negative effects of contamination by methylmercury and other molecules on loggerhead turtle populations.

2. Materials and Methods

In this study, loggerhead RBCs were used as an in vitro model for determining the oxidative stress of acute MeHg concentrations. The RBCs were subjected to a pre-adaptation process to the supplemented MEM culture medium. The LC_{50} of RBCs to methylmercury was determined at concentrations from 0.5 to 100 mg L^{-1} for 96 h of incubation. The determined LC_{50} was taken into account to assess oxidative stress. The RBCs were subjected to acute concentrations of MeHg (0, 1, and 5 mg L^{-1}) and incubated at 30 °C. It was evaluated whether the SOD, GST, and MDA activities at 6 and 12 h after MeHg exposure can be used as biomarkers of environmental contamination and oxidative stress in loggerhead turtles (Figure 1).

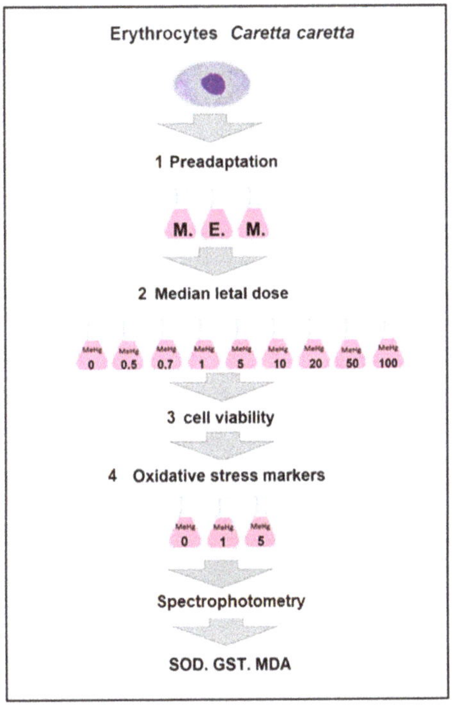

Figure 1. Experimental design. The loggerhead sea turtle erythrocytes were subjected to 1. Pre-adaptation in supplemented Minimum Essential Medium (MEM) medium, 2. Determination of LC_{50} in nine doses of methyl mercury, 3. Cell viability determination and 4. The erythrocytes were subjected to acute doses of methyl mercury and the activities of superoxide dismutase (SOD) and glutathione-s-transferase (GST) enzymes, and the Malonaldehyde (MDA) concentration were measured for spectrophotometry.

2.1. Study Area, Collection and Analysis of Samples

The peripheral blood samples were collected from five inmature, apparently healthy, captive-reared pre-juvenile loggerhead sea turtles. The turtles were maintained at ambient temperature (average 30 °C, minimum 27 °C) in an outdoor seawater pool at the El Rodadero Aquarium in Santa Marta,

Colombia (11°13′1.09″ N, 74°14′13.75″ W), which was characterized by a semi-arid climate (Figure 2). The samples were obtained under a research permit that was granted by the Ministry of Environment and Territorial Development (#24 of 22 June 2012) and Contract for Access to Genetic Resources (#64 of 23 April 2013). We also have a permit for the collection of samples of Colombian biodiversity, issued to the Jorge Tadeo Lozano University (Resolution 1271 of 23 October 2014, IDB0401 File). The turtles were visually assessed for any injuries to the flippers and carapace. Turtles were considered pre-juveniles due to their development status and Straight Carapace Length (SCL) <34 cm [36,37]. Turtles were marked, measured, and weighed (Table 1). Blood samples were extracted from the dorsal region of the cervical sinus, in accordance to previously described methodology by Dutton [38]. For the collection of peripheral blood, 4 mL sterile syringes and 2 mL tubes with sodium heparin were used. Standard blood tests were performed by Microvet Laboratories, Bogotá D. C., prior to the bioassay in this study, in order to assess the health status of the turtles (Table 1). The total mercury (Hg-T) in RBCs and seawater was analyzed by atomic absorption spectrophotometry at 254 nm (DMA-80 evo, Milestone-24010 Sorisole (BG), Italy). The 100 μL of sample, without any pre-treatment, were introduced into a sample boat, dried through a prescribed method in the DMA-80, and burned at 800 °C. Devolatilized elementary Hg (Hg vapors) was collected on a gold amalgamation trap and subsequently desorbed for quantification. DMA-80 reaches a detection limit as low as 0.001 nanograms of mercury. Analyses of duplicates and standard reference material were performed to control accuracy and precision. This quantification was carried out at Toxicology Laboratory, COLCAN, and Bogotá D. C.

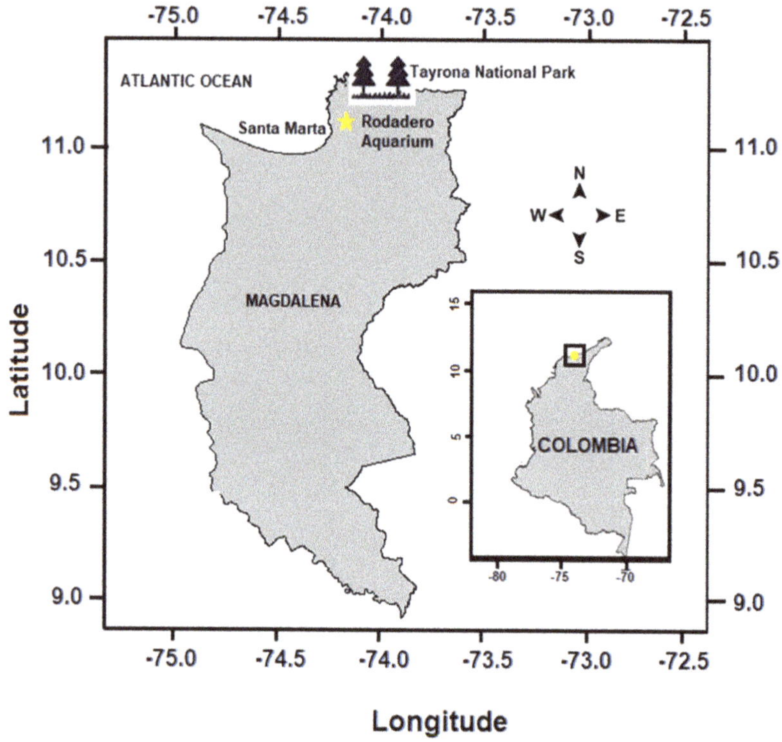

Figure 2. Location of El Rodadero Aquarium (11°13′01″ N 74°14′14″ W), Santa Marta (Colombia), site for peripheral blood sampling site of the five individuals of loggerhead sea turtles.

2.2. Experimental Procedures

2.2.1. Isolation and Pre-Adaptation of Loggerhead Sea Turtles Erythrocytes

All of the experiments were carried out in the laboratory of genetics and molecular biology of the Jorge Tadeo Lozano University. The RBCs of peripheral blood were isolated by centrifugation at 1500 rpm for 2 min at 4 °C. Plasma and buffy coat were removed. The RBCs packed in pellet were washed five times with phosphate buffered saline (PBS, 0.9% NaCl in 10 mM sodium phosphate buffer, pH 7.4) and suspended to prepare a 2% (v/v) cell suspension in Minimum Essential Medium (MEM) (Gibco, Grand Island, NY, USA), supplemented with sodium carbonate, 13% fetal bovine serum (FBS), and 1% antibiotics and antimycotics (100 mg/L penicillin, 10 mg/L streptomycin) (MEM-S). In this culture medium, the RBCs were pre-adapted for 96 h, incubated at 30 °C, and then used in the following bioassays.

2.2.2. Determination of LC_{50} of Loggerhead Sea Turtles Erythrocytes

Solutions of methylmercury (II) chloride CH_3HgCl (Sigma–Aldrich, St. Louis, MO, USA) were prepared in PBS buffer at concentrations of 0.5, 0.7, 1, 5, 10, 20, 50 y 100 mg L^{-1} to determine the LC_{50}. An experimental unit was used for each concentration, plus an additional control group, each having three replicates. Each replica was served in 1.5 mL Eppendorf tubes containing 2.6×10^6 RBCs/mL dissolved in 180 μL of MEM-S and 5 μL of the appropriate MeHg solution. The cells were incubated at 30 °C for 96 h. Distilled water was added to the control group and the bioassay was repeated three times. At 96 h, aliquots of 10 μL of RBCs were mixed with 10 μL of trypan blue solution (Sigma–Aldrich; Merck KGaA). Living and dead cells were counted in a Neubauer counting chamber (Zeiss, Oberkochen, Germany) while using the Leica DM750 microscope (Leica Microsystems, Heerbrugg, Switzerland) with 40× magnification. LC_{50} was calculated through a Probit analysis [39] using the "MASS" package [40] that was included within the free RWizard software [41].

2.2.3. Determination of Cell Viability of Loggerhead Sea Turtles RBCs

Cell viability was evaluated at 0, 1, and 5 mg L^{-1} of MeHg. These doses were selected based on the results of the LC_{50} test, which were also supported by previously published results [42]. The cell viability test was performed in triplicate, then, 2.6×10^6 RBCs/mL dissolved in 180 μL of MEM-S and 5 μL of the appropriate MeHg were incubated at 30 °C for 6 and 12 h of exposure. Subsequently, 10 μL of 0.4% (v/v) trypan blue solution (Sigma-Aldrich; Merck KGaA) and 10 μL of RBCs suspension subjected to each of the MeHg concentrations were mixed in a polypropylene microtube and was incubated for 5 min at room temperature. 10 μL of the Trypan blue/RBC cell mixture were then placed on a Neubauer counting chamber (Zeiss, Oberkochen, Germany) and, subsequently, stained (non-viable) and unstained cells (viable) were counted while using a 40× magnification microscope (Leica DM750 microscope, Leica Microsystems, Heerbrugg, Switzerland).

2.3. Oxidative Stress of Loggerhead Sea Turtles Erythrocytes

SOD and GST, 2×10^6 RBCs/mL of loggerhead turtles exposed and not exposed to MeHg (0, 1 and 5 mg L^{-1}) were used to assess oxidative stress markers. These were lysed with four volumes of ice-cold HPLC, and then centrifugated at 10,000 *g* for 15 min at 4 °C. The supernatant (RBCs lysate) was collected for assaying and stored at −80 °C. For the MDA assay, 2×10^7 RBCs/mL of loggerhead turtles exposed and not exposed to MeHg (0, 1, and 5 mg L^{-1}) were used. The samples were centrifugated at 2500 rpm for 5 min at 4 °C, and then resuspended in 1 mL of MEM. RBCs were collected and stored at −80 °C until use.

The activity of Superoxide Dismutase (SOD) and Glutathione S-Transferases (GST), and the levels of Malonaldehyde (MDA) were determined in RBCs lysate using commercial kits (Superoxide Dismutase Assay Kit [Item No. 706002], Glutathione S-Transferase Assay Kit [Item No. 703302], and TBARS assay kit [Item No. 10009055], Cayman Chemical Co, Ann Arbor, MI, USA), following the manufacturer's protocol for each Kit.

The Superoxide Dismutase (SOD) activity measuring kit is based on a method in which xanthine and xanthine oxidase react with tetezolium salt (2-(-4-iodophenyl)-3-(4-nitrophenol)-5-phenyl tetrazolium chloride), generating superoxide radicals, which then react and form formazan red. The SOD activity was then measured by the degree of inhibition of this reaction. In a 96-well microplate, 10 μL of the Standard SOD enzyme (included in the kit, in AG wells) were served. In the rest of the microplate, 10 μL of previously exposed and unexposed lysed RBCs (0, 1, and 5 mg L^{-1} of MeHg) were served (in triplicate), then, 200 μL of the diluted radical detector were added to all wells. To start the reaction, 20 μL of diluted xanthine oxidase were added to all wells, both standard and samples. Xanthine oxidase enzyme was not added for the control and the recorded readings were used to make the absorbance correction once the measurements were made. The plate was shaken for 30 min at room temperature and then the absorbance at 440–460 nm was measured in a plate reader (Elisa reader Expert 96 ASYS Hitech, Eugendorf, Austria). The average absorbance of each standard and sample was calculated and then corrected, subtracting the absorbance recorded in the control. The linearized rate (LR) of all standards was calculated, first dividing the absorbance of standard A by itself and then that of the other BG standards dividing the value recorded in A, by those that were recorded in each of the BG standards. This means LR for Std A = Abs Std A/Abs Std A; LR for Std B = Abs Std A/Abs Std B, and so on. The linear standard SOD rate (LR) was plotted against the final SOD activity (U/mL).

The SOD activity of the samples was calculated while using the equation that was obtained from the linear regression of the standard curve, substituting the linearized rate (LR) for each sample. A unit of SOD is defined as the amount of enzyme that is required to exhibit a 50% dismutation of the superoxide radical. SOD activity is standardized while using the coupled cytochrome c and xanthine oxidase assay. The results were presented in international units per milliliter (U/mL).

$$OD\left(\frac{U}{mL}\right) = \left[\left(sample\ LR - y - \frac{intercept}{slope}\right) \times \left(\frac{0.23\ mL}{0.01\ mL}\right)\right] \times Sample\ dilution \quad (1)$$

The GST enzyme was quantified in triplicate in RBCs lysate. 150 μL of assay buffer, 20 μL of glutathione, and 20 μL of each of the previously exposed and unexposed lysed RBC samples (0, 1, and 5 mg L^{-1} MeHg) were served in a 96-well microplate. As a negative control, 170 μL of test buffer and 20 μL of glutathione were used, and as a positive control, 150 μL of test buffer, 20 μL of glutathione, and 20 μL of reconstituted GST. The reaction was started by adding 10 μL of CDNB (included in the kit) to all wells and gently shaking the microplate. GST activity was read at 340 nm on a plate reader (Elisa reader Expert 96 ASYS Hitech, Eugendorf, Austria). The addition of the GST produced an increase of between 0.012 and 0.064/min. The change in absorbance over time was calculated while using the following equation, (where A represents the absorbance at 340 nm):

$$\frac{\Delta A}{min} = \frac{A\ (Time2) - A(Time1)}{Time2(min) - Time1(min)} \quad (2)$$

GST activity was calculated using the following formula:

$$ST\ Activity = \frac{\frac{\Delta A}{min}}{\frac{1}{0.00503\ \mu M}} \times \left(\frac{0.2\ mL}{0.02\ mL}\right) \times Sample\ dilution = \frac{\frac{nmol}{min}}{mL} \quad (3)$$

The reaction rate can be determined while using the CDNB extinction coefficient of 0.00503 μM^{-1}. One unit of the enzyme will be conjugated to 1.0 mMol of CDNB with reduced glutathione for one minute at 25 °C. The actual extinction coefficient para CDNB at 340 nm is 0.0096 μM^{-1} cm^{-1} [43]. This value has been adjusted for the pathlenght of the solution of the well (0.524 cm).

The MDA concentration in RBCs was quantified in duplicate using the RBCs collected previously and stored at −80 °C. RBCs were sonicated at 36% amplitude with three pulses of 20 s each, using a sonicator (Sonics Vibra Cell, 750 W, Newtown, CT 06470, USA). RBCs thatwere resuspended in MEM and malonaldehyde standards (100 µL MDA, included in the kit) were mixed in 5 mL falcon tubes with 100 µL of SDS and 4 mL of the color reagent (included in the kit). All tubes were placed in a serological bath for 1 h at 100 °C, then incubated on ice for 10 min to stop the reaction. They were then spun at 1600 rpm for 10 min at 4 °C. 160 µL aliquots of each of the treatments and standards were added to a 96-well microplate. MDA measurement was performed spectrophotometrically at 530 nm, on a plate reader (Elisa Expert 96 reader (ASYS Hitech, Eugendorf, Austria). The measurement of MDA concentration was reported in µM.

Absorbances in all cases were corrected, as follows: (1) the average absorbance of each sample and the standard sample was calculated, (2) the absorbance value was subtracted from the standard A (0 µM) of itself and the other values (both standards and samples, and the corrected absorbance were found), (3) the corrected absorbance of the standard samples was graphed as a function of the MDA concentration, and (4) the MDA values for each sample of the standard curve were calculated using the equation:

$$MDA(\mu M) = \frac{(corrected\ absorbance) - (y - intercept)}{Slope\ of\ the\ line} \quad (4)$$

2.4. Data Analysis

Statistical analyzes for biochemical biomarker data (SOD, MDA, and GST concentration) in controls and treatments were subjected to a normality test and the homogeneity variance test using the Shapiro Wilk, Kolmogorov–Smirnov, and Levene tests, respectively. Significant differences between treatments and control were determined by an ANOVA test and then by multiple comparisons between treatments using the Tukey test. The sample size did not allow for stipulating significant differences between the treatments. The correlation coefficients of Pearson, Spearman, and Kendal were calculated to determine whether there is a correlation between the concentrations of MeHg to which the erythrocytes were exposed and the observed levels of GST, SOD and MDA at 6 and 12 h of exposure. To carry out the analyzes and figures, the R program was used [44] with RWizard software [41].

3. Results

3.1. Blood Count and Health Status of Loggerhead Turtles

The sampled individuals' sex could not be determined because of their immature development state. They had an average weight of 2.78 ± 0.29 kg, an average straight carapace length (SCL) of 29.4 ± 2.41 cm, and an average curve carapace length (CCL) of 26.2 ± 2.71 cm (Table 1). With these SCL, turtles are considered to be pre-juveniles [36,37]. The blood count was conducted prior to the MeHg bioassay in order to determine the health status of the turtles. All of the specimens presented an apparently good state of health that was corroborated by all of the obtained blood parameters (Table 2). The percentages of lymphocyte count, mean corpuscular volume, and the percentage of lymphocytes, eosinophils, and monocytes were similar to those previously reported [34]. The variations in concentration of the different parameters were within the accepted levels [34,35]. Nonetheless, the blood parameters between the different species of sea turtles, and between free-living or captive individuals of the same species are different and, in some cases, highly variable, depending on their geographical distribution, [4,45,46], as numerous additional intrinsic and extrinsic factors can impact blood composition.

Table 1. Body size of the f specimens of loggerhead sea turtles sampled for erythrocytes, identification number, live weight, Straight Carapace Length (SCL), and Curved Cap Width (ACL).

ID of Turtles	Weight (Kg)	SCL (cm)	ACL (cm)
UJTL-CC1	3.0	31	28
UJTL-CC3	2.9	30	28
UJTL-CC5	2.6	29	24
UJTL-CC7	2.3	25	22
UJTL-CC8	3.1	32	29

Table 2. Blood parameters of captive loggerhead sea turtles from El Rodadero Aquarium, Santa Marta. SD = standard deviation (n = 5). Reference values taken from Yang et al., and Rouselet et al. [47,48]. M.C.V.: medium corpuscular volume; H.C.M.: amount of hemoglobin per red blood cell; C.M.H.C.: mean corpuscular hemoglobin concentration.

Variable	Mean	SD	Min	Max	Reference	Unit
HEMOGRAM						
ERYTHROCYTES	1.42	0.13	1.21	1.61	1.60–70.0	10^5 Cells/μL
HEMOGLOBIN	3	0.59	2.1	3.7	3.9–9.5	g/L
HEMATOCRIT	6.98	1.20	5	8.4	9.5–36.0	(%)
M.V.C	488.5	49.97	413.2	560	257.1–1188	Femtoliters
H.C.M.	210.2	30.37	174	247	157.7–216.3	Picograms
C.M.H.C.	43.2	5.91	37	54	24.1–32.4	(g%)
PLASMATIC PROTEIN	9.74	1.23	8.5	12	1.2–5.0	g/L
ALBUMIN	6.72	0.89	5.4	8	0.8–2.2	g/L
GLOBULINS	3.02	1.44	1.8	5.6	0.2–3.3	g/L
THROMBOCYTES	24	12.65	3	42		Cells/μL
LEUCOCYTES	1.83	0.14	1.59	1.99	1.760–23.60	10^3 Cells/μL
HETEROPHILE	42	2.37	38	45	14.8-91.6	(%)
BASOPHILS	3.6	0.8	3	5	0–16	(%)
EOSINOPHILS	3	1.09	2	5	0-13.7	(%)
LYMPHOCYTES	45	3.03	41	49	17.2-90	(%)
MONOCYTES	3.4	1.85	1	6	0-16.1	(%)
AZUROPHILES	3	1.41	1	5	1-8	(%)
ABSOLUTE PARAMETERS						
LYMPHOCYTES	820	50	747	892	429.8–11034.3	Cells/μL
HETEROPHILE	790	60	692	856	1308.1–8400.9	Cells/μL
EOSINOPHILS	50	20	32	98	13.6–2580.2	Cells/μL
MONOCYTES	60	40	18	117	7.2–2796.6	Cells/μL
BASOPHILS	60	20	0.048	0.1	0.064–0.928	10^3Cells/μL
AZUROPHILES	60	30	15.9	99.5	78.4–729.9	Cells/μL

3.2. *Quantification of Hg in Erythrocytes of Loggerhead Turtles*

The concentration of Hg in seawater samples and in 100 μL samples of RBCs was determined by atomic adsorption (see details above). The background Hg-T concentrations in control seawater was 0.0032 ± 0.001 mg L^{-1} (0.00323 ppm), whereas the background Hg-T concentrations in RBCs (before the bioassay) were 0.031 ± 0.011 mg L^{-1} on average. After the bioassay, at 12 h, the RBCs exposed to concentration of 1 mg L^{-1} (1 ppm) had an average Hg-T concentration of 0.31 ± 0.10 mg L^{-1} and those exposed to 5 mg L^{-1} (5 ppm) contained 2.18 ± 0.77 mg L^{-1}. These results demonstrate the high affinity of RBCs for Hg accumulation, due to their great capacity to retain it in cell membranes [49].

3.3. *Lethal Concentration (LC_{50}) and Cell Viability of Erythrocytes of Loggerhead Turtles*

Loggerhead RBCs that were exposed to 50 and 100 mg L^{-1} of MeHg for 96 h had 100% cell mortality (Figure 3). For all other exposure concentrations, cell viability ranged from 90% to 20%. From these data, the LC_{50} was obtained with a value of 8.32 mg L^{-1} (8.32 ppm). Based on these results, it was decided to perform the bioassays for the determination of oxidative stress in the RBCs of loggerhead turtles at 0, 1, and 5 mg L^{-1} of MeHg, and for incubation durations of 6 and 12 h.

In the bioassay of cell viability, the RBCs that were exposed to 0, 1, and 5 mg L^{-1} of MeHg for 12 h showed 100% viability.

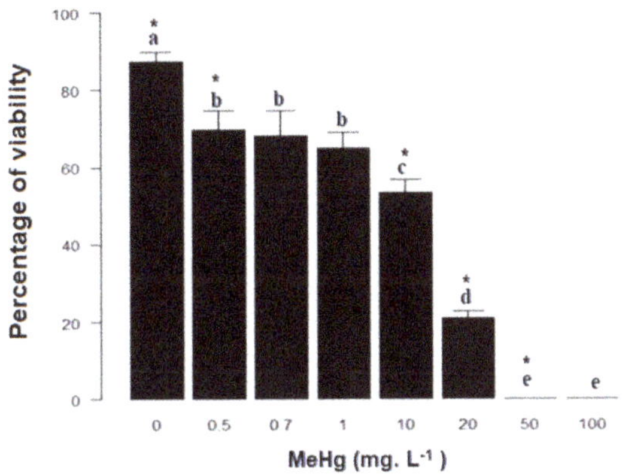

Figure 3. Cell viability of erythrocytes of loggerhead sea turtle against different concentrations of methylmercury, used for calculating lethal concentration (LC_{50}). Error bars indicate mean and standard deviation. Treatments with significant differences (ANOVA, $p < 0.05$) indicated by asterisk (*). Treatments with non-significant differences indicated by same letter (a, b, c, d, e).

3.4. *Bioassay and Oxidative Stress*

The SOD activity presented variations in each of the different treatments. At 6 h, the SOD activity of RBCs exposed to 1 mg L^{-1} of MeHg was slightly more than double that in controls (1.56 ± 0.58 vs. 0.72 ± 0.24 U/mL of RBCs), and slightly greater than at 5 mg L^{-1} MeHg (1.20 ± 0.52 U/mL), which, in turn, was higher than in the control, demonstrating that RBCs at 1 and 5 mg L^{-1} of MeHg experienced oxidative stress (Figure 4A). After 12 h of exposure, SOD activity was higher in RBCs that were exposed to 5 mg L^{-1} MeHg (1.39 ± 0.45 U/mL) than in controls (0.82 ± 0.24 U/mL) and those RBCs exposed to 1 mg L^{-1} of MeHg (1.08 ± 0.25 U/mL), again indicating oxidative stress. There were no statistically significant differences between RBCs' SOD activity for the two MeHg exposure concentrations (1 and 5 mg L^{-1}) for either of the two exposure durations (6 and 12 h), no correlation between the SOD activity and MeHg concentration to which the RBCs were exposed after 6 h of exposure (in all cases $p > 0.05$).

The glutathione transferase activity of loggerhead turtle RBCs at 6 and 12 h of exposure showed variations in cells that were exposed to 1 and 5 mg L^{-1} of MeHg with respect to the control. At 6 h, the GST activity was higher for the 1 and 5 mg L^{-1} MeHg treatments as compared to the control (12.72 ± 5.41 and 11.93 ± 9.54 vs. 6.76 ± 4.93 nmol/min/mL of RBCs, respectively, demonstrating high response to the concentration and exposure time to MeHg. However, after 12 h of MeHg exposure, the GST activity was lower in RBCs exposed to 5 mg L^{-1} MeHg when compared to those exposed to 1 mg L^{-1} MeHg and to the control (5.96 ± 6.36 vs. 9.54 ± 8.91 and 6.92 ± 5.16 nmol/min/mL of RBCs), respectively. Nevertheless, the observed differences between MeHg treatments and the control were not statistically significant for any exposure concentrations or time point (in all cases $p > 0.05$) (Figure 4B).

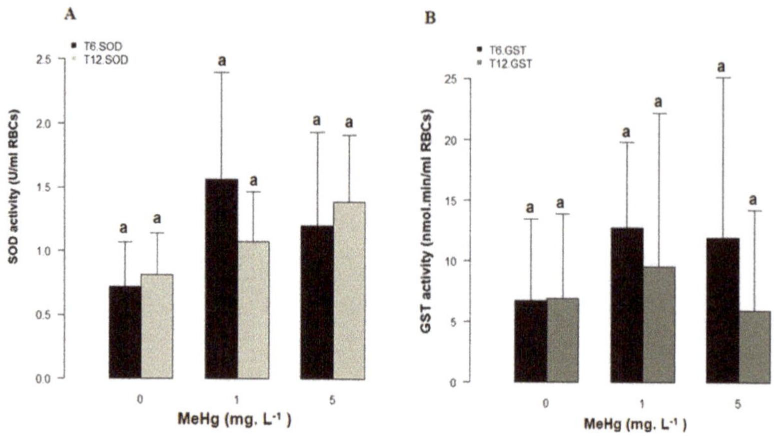

Figure 4. Activities of (A) Superoxide dismutase (SOD) and (B) glutathione S-transferase (GST) in erythrocytes of loggerhead turtles exposed to methylmercury (MeHg) at 0, 1 and 5 mg L^{-1} for 6 and 12 h (mean ± standard error of the mean, (n = 5). Differences between treatments were statistically non-significant in all cases ($p > 0.05$).

The malondialdehyde (MDA) levels increased with respect to the control with increasing MeHg concentration and exposure time. After 6 h, the MDA concentrations were 38.31 ± 20.92 and 46.65 ± 13.42 μM for 1 and 5 mg L^{-1} MeHg, respectively, when compared to 27.58 ± 0.42 μM in the controls. After 12 h, the MDA concentrations increased to 49.15 ± 15.75 and 56.38 ± 13.97 μM of MDA for 1 and 5 mgL^{-1} MeHg, respectively, whereas MDA in controls remained virtually unchanged (28.91 ± 9.35 μM, Figure 5). While the observed differences between treatments were not considered to be statistically significant (ANOVA, $p > 0.05$), a positive correlation was observed between the MDA concentrations and MeHg concentrations after 12 h of exposure time (r Spearman = 0.56, $p < 0.05$, Figure 6). This result is preliminary; therefore, it is necessary to carry out evaluations with a larger sample to improve the statistical significance of this analysis and verify this possible correlation.

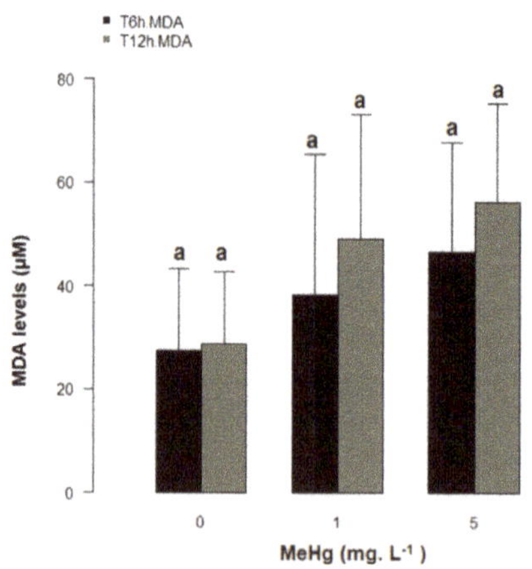

Figure 5. Malondialdehyde (MDA) levels in control (untreated erythrocytes) and in erythrocytes exposed to MeHg at 1 and 5 mg L^{-1} for 6 and 12 h (mean ± standard error of the mean, n = 5). Differences between treatments were non-significant in all cases ($p > 0.05$).

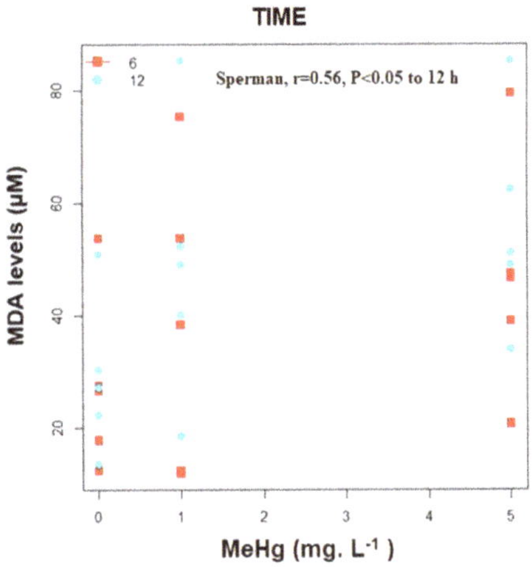

Figure 6. Spearman's correlation between the variables MeHg concentration, exposure time and MDA levels.

4. Discussion

The present study is the first report on oxidative stress of RBCs of captive loggerhead turtles (*Caretta caretta*) that were exposed to methylmercury in bioassays. RBCs of sea turtles have not been used to perform in vitro bioassays until today. This study showed RBCs to be a sensitive, simple, and inexpensive model for performing bioassays for ecotoxicological evaluation of heavy metals, such as mercury and, potentially, other environmental pollutants. Furthermore, SOD, GST, and MDA were sensitive biomarkers of environmental pollution and oxidative stress in loggerhead turtles.

The blood count was conducted prior to the MeHg bioassay, in order to show the health status of the turtles. Plasma and hematological biochemistry analytes vary significantly between loggerhead turtle populations in the wild, due to environmental factors, temperature, type of diet, sex, size, pH, salinity, water quality, population density, development stage, maturity, weight, and season [50,51]. Additionally, the hematological response to man-made influences must be taken in consideration when reviewing previously published data. For these reasons, it is important to establish population-specific reference information [52]. In this study, all of the specimens presented an apparently good health showed by the blood parameters obtained (Table 2). The percentages of all white blood cells were within the previously described ranges for loggerhead turtles (Figure 7, [47]). Additionally, mean corpuscular volume, albumin, and plasma protein were similar to those reported previously [47,48,51–53]. The variations in concentration of the different parameters were within the accepted levels [53,54]. A simple analysis that is presented in Figure 7 shows that the percentages of white cells: heterophile, eosinophils, limphocythes, monocytes, and azurophiles are within the reference ranges [47] that are present in loggerhead turtles with an apparent good health. However, as we have seen, the size and weight of turtles are important variables that produce large variations in the concentrations of the different hematological analytes. In the study that was carried out by Yang et al. (2019) [47], the 191 analyzed individuals are 2.5 times larger than the individuals in our study according to the SCL range (54.8–100.8 cm vs. 25–32 cm). For this reason, the percentages of the heterophile and limphocythes in this study are approximately half of the maximum values (91.6% and 90%, respectively). In the same way, the eosinophils, monocytes, and azurophiles found are four, five, and three times under the maximum value (13.7%, 16.1%, and 8%, respectively). All of the cell types exceed the minimum range to present them within the normal ranges. These differences are apparently a result of the weight and size of the turtles, but, as described above, it might be due to other variables, such as captivity, feeding, and/or stress [50,51].

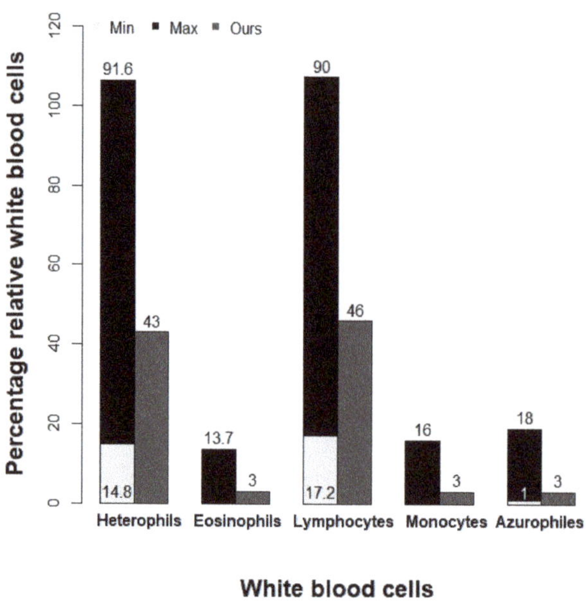

Figure 7. Comparison between the percentages relative to white cells in the blood counts for the loggerhead turtles (*Caretta caretta*) (ours) and the minimum and maximum ranges published by Yang et al. (2019).

This study used captive loggerhead turtles. For this reason, the background Hg-T in water and in turtle RBCs were determined prior to bioassays (0.0032 ± 0.001 mg L^{-1} and 0.031 ± 0.011 mg L^{-1}, respectively). The 10-fold concentration of baseline Hg-T in RBCs demonstrates how feeding is an important factor for the bioaccumulation and biomagnification of this metal. The diet of loggerhead turtles in their natural habitat is composed of cnidarians, gastropods, copepods, and other crustaceans, polychaetes, cephalopods, fish, algae, plant remains, fish eggs, and hydrozoans [55]. In this study, captive donor loggerhead turtles from Santa Martha Aquarium were fed with fish of species as machuelo (*Opisthonema oglinum*), macabí (*Elops saurus*) and picuda (*Sphyraena ensis*). It is presumed that fish diet had Hg concentrations like those that were previously reported for machuelo (*Opisthonema oglinum*) (0.11 μg L^{-1}) [56].

In the present study, the average total mercury (Hg-T in RBCs) concentration of 0.031 ± 0.011 mg L^{-1} was obtained prior to bioassay and, after 12 h of bioassay with erythrocytes at 1 mg L^{-1} MeHg, the concentration of Hg-T increased approximately 10 times (0.309 ± 0.10 mg L^{-1}). At 5 mg L^{-1} of MeHg, the Hg-T concentrations increased to 2.183 ± 0.77 mg L^{-1}, being 70 times higher than controls and seven times higher than RBCs that were subjected to 1 mg L^{-1} of MeHg.

Various studies have shown that the different dissolved forms of mercury have a high affinity for the sulfhydryl groups present in the cells, and that in their methylated state it has an affinity of over 90% for human RBCs, where it adheres in less than an hour [22,57,58]. Studies of RBCs of mammalian species have also shown that MeHg is evenly distributed between intra and extracellular hemoglobin [59,60]. In this study, these results demonstrate the high affinity of RBCs for Hg and their great capacity to retain it inside of the cell.

Moreover, the MeHg median lethal concentration (LC_{50}) of RBCs for loggerhead turtles exposed in vitro was determined in this study, eight MeHg doses ranging from 0.5–100 mg L^{-1}, in addition to seawater controls. At the two highest concentrations, there was 100% cell mortality (Figure 3). In the control group, mortality was 13% and it increased with the MeHg concentration. The LC_{50} for loggerhead turtle RBCs was 8.32 mg L^{-1} after 96 h of incubation. To date, there are no published studies on LC_{50} of MeHg on loggerhead sea turtle RBCs, so there are no literature values to compare these results with. Recently, the use of RBCs was reported in turtle studies of the red eared sliders species (*Trachemys callirostris*) to assess the genotoxic effect of environmental pollutants [61]. Cocci et al. [62] described the use of primary erythrocyte cell cultures as in vitro models for evaluating the effects of different endocrine disrupting chemicals. Finlayson et al. [55] established primary cell cultures from the intestine, heart, liver, ovary, and skin of green sea turtles (*Chelonia mydas*), and skin for loggerhead turtles and determined the mean effective concentration (EC_{50}) with an average for all cell types of 18,4 μM for Hg^{+2}, being skin tissue the most sensible. In this study, LC_{50} was found to be almost double this value (33.05 μM). They found significant differences between the different cell types evaluated. In this study, other cell types, RBCs, were evaluated in loggerhead turtles, also showing great differences. The differences in response to Hg contamination between cells and between different species are appreciable [63–65]. Cytotoxicity is generally due to non-specific molecular and cellular effects [55]. Some pollutants, such as Hg, have higher accumulation values in loggerhead turtles [66,67]. Loggerhead sea turtles' RBCs appear to have a higher sensitivity and affinity for Hg than skin cells and other tissues (as described above), which would produce the big difference between these two results. RBCs could represent a better in vitro model for toxicological studies than primary cell cultures according to this result, and due to the high sensitivity of the RBCs presented here. However, primary cell cultures can be used to specifically study the response of organs to environmental pollutants. Viability studies that were conducted in RBCs in marine fish species as *Sparus aurata* and *Dicentrarchus labrax* have reported LC_{50} of 5.25 and 7.25 mg L^{-1}, respectively, in RBCs that were exposed to MeHg for 24 h [49]. This result is similar to the present study, although the exposure times are different, showing that the RBCs of the turtles are more resistant. We found that loggerhead turtle RBCs can survive at 4 °C for about eight weeks when stored with plasma. The determination of the LC_{50} allowed for selecting the appropriate concentrations of MeHg to be used in the bioassay (0, 1, and 5 mg L^{-1}), so that the majority of RBCs remained alive until the end of the bioassay, as corroborated by cell viability close to 100%.

Studies on oxidative damage to wild animals from an ecological perspective have been conducted over the past 20 years [68], but knowledge on oxidative stress in sea turtles is scarce [1,55,69,70]. Further, the toxic level of Hg and others metals and their short- and long-term adverse effects are still unknown for sea turtles [1,55,71].

RBCs have an antioxidant defense system that is composed of the enzymes superoxide dismutase (SOD), catalase (CAT), and glutathione peroxidase (GPx) and molecules, such as reduced glutathione (GSH), tierrodoxins, alpha tocopherol, beta carotene, and vitamin C [72]. The exposure of RBCs to methylmercury induces oxidative stress, which destabilizes the lipid bilayer of the cell membrane and causes morphological and physiological changes in the cell [63,73]. RBCs increase or decrease the expression levels of catalase, superoxide dismutase, and glutathione enzymes in order to counteract stress. These enzymes have been monitored in sea turtles in blood and liver tissues [74,75] to determine the correlation between the bioaccumulation of metals and oxidative stress and, in this way, understand the possible mechanisms that sea turtles use to balance themselves. There are no reports of in vitro bioassays of RBCs of sea turtles that were exposed to MeHg where oxidative stress is evaluated.

In this study, where loggerhead RBCs exposed in concentrations of 5 mg L^{-1} of MeHg, cells presented lipid peroxidation and reduction in the enzymatic activity of SOD and GST after 6 h of exposure. It seems that high concentrations of MeHg inhibit the expression of enzymes and, therefore, their activity [76–78]. These results are similar to those that were reported by Berntssen et al. [76] in brain, liver, and kidney tissues of fish that were exposed to 5 and 10 mg L^{-1} MeHg. Greater susceptibility of brain tissue to oxidative stress was reported, which was represented by increased levels of lipid peroxidation (MDA) and ROS formation, and parallel reduction in the activity of SOD and GST enzymes in brain cells with increasing MeHg concentration (up to 10 mg L^{-1} MeHg).

The increase of the enzymes Cu, Zn-SOD, and Mn-SOD or γ-glutamylcysteine synthetase is the first adaptive response that cells present when exposed to mercury compounds [79]. However, studies in mice that were exposed to Hg and MeHg have shown that doses of these compounds reduce glutation and SOD levels, respectively, in liver and brain tissues [80,81]. These experimental results agree with the results shown in this study, and they would indicate that loggerhead turtle RBCs are as susceptible to oxidative stress that is caused by MeHg exposure as the brain cells of the fish [76], mouse, and insect neuronal and liver cells [79], since these same trends were observed for the activity of the SOD and GST enzymes, and lipid peroxidation.

The SOD and GST enzymes make up the first line of cellular defense against oxidative stress, neutralizing oxygen and nitrogen reactive species and preventing lipid peroxidation [20]. When the levels of these enzymes and other non-enzymatic antioxidants, such as cysteine, which participates in the uptake and release of mercury [79], are not intracellularly sufficient or their production levels reach maximum ceilings, their activity is reduced and metals and free radicals that could not be extracellularly expelled cause the degradation of PUFAS membrane and, therefore, lipid peroxidation [60].

MeHg is known to generate lipid peroxidation [82]. Studies have shown that long-lived species have erythrocyte cell membranes that are more resistant to lipid peroxidation, which might vary with the age of the individual and, therefore, be a direct predictor of survival [83]. In the present study, it was observed that the mean levels of MDA increased at 6 and 12 h of exposure, even though observed differences between exposure concentrations were not statistically significant due to the high variance of results. Apparently, the low number of individuals (n = 5) could explain this result. A positive correlation was observed between the concentration of MeHg and the levels of MDA, $r = 0.56$, $p < 0.05$ after 12 h of exposure (Spearman). Studies with a higher n should be performed to verify this possible correlation due to the low number of samples used.

In this study, we found that an in vitro increase in MeHg concentration caused a corresponding increase in MDA concentration (LPO), as well as decreases in the activities of SOD and GST, result which is in agreement with previous studies [76–78,84]. Nonetheless, it is important to mention that there are other sensitive enzymes that respond to MeHg that should be studied as biomarkers of toxicity to mercury compounds. Peroxiredoxin and thioredoxins are found in all tissues and are involved in multiple functions that are associated with cellular maintenance and survival, including protein repair, regulation of the cell cycle, and cell signaling [85]. These enzymes could function in host defense against oxidative stress, modulate immune response, and control intracellular peroxide levels [86,87], as well as participate in the antioxidant defense system [88,89].

Furthermore, it was reported for Chinese soft-shelled turtle (*Pelodiscus sinensis*) that peroxiredoxins was implicated defense against microbial pathogens and oxidants and would provide information of functional mechanism of immunity [90]. Studies that were carried out with these enzymes in human cell lines have reported that mercury and MeHg bind to their thiol groups (–SH), decreasing activity and causing inactivation in most cases [91–93]. Peroxiredoxins and thioredoxins are present in blood and, thus, are very important for living beings. They are very sensitive to oxidation [94,95], but, thanks to their plasticity, they have the ability of making post-translational modifications that prevent them from oxidizing, maintaining the homeostatic balance of cells [96]. Thioredoxins and peroxiredoxins function structurally and they have been barely studied in turtles [90]. Some of these studies have focused on evaluating how brevotoxins affect thioredoxin, GST, and SOD levels [97]. Moreover, turtles' regulation of intracellular levels of GST and peroxiredoxins has been studied under oxidative stress caused by anoxia, freezing, and exposure to lipopolysaccharides [98–100]. Overall, studies to date conclude that peroxiredoxins and thioredoxins are important targets for the compounds of mercury and their inhibition is likely to explain the results beyond Hg-induced toxicity [101]. For this reason, the evaluation in future studies of TrxR and Trx in blood and plasma could provide a useful tool for the prediction of adverse effects that are induced by mercury compounds.

Additionally, the results indicate that mercury can inhibit the activities of many proteins/enzymes that are involved in combating oxidative stress in cells, including thioredoxin (Trx) (MeHg and $HgCl_2$), glutaredoxin (Grx) (MeHg) [91,102], glutathione peroxidase (Gpx) (MeHg) [103], glutathione reductase (GR) (MeHg and $HgCl_2$), thioredoxin reductase (TrxR) (MeHg and $HgCl_2$) [91,104], superoxide dismutase (SOD) (MeHg and $HgCl_2$) [105], neuronal nitric oxide synthase (nNOS) (MeHg) [105], and Keap1 [106]. In addition, mercury can also alter the ratio of GSH/glutathione disulfide (GSSH) either by directly binding to GSH, depleting GSH, or decreasing the levels of Gpx [102,107,108].

For species, such as sea turtles, where obtaining in vivo effect values is an impossibility, in vitro data provides a useful opportunity for conduct research into chemical risk to these species. As the field of sea turtle toxicology continues to expand, the results from studies such as this provide crucial information for the future experiments and guiding management and conservation plans [55].

It is necessary to improve the state of knowledge about the variability of indicators of oxidative stress in loggerhead turtles and their relationship to concentrations of MeHg. For this purpose, the use of erythrocytes as biomarkers of environmental pollution is a promising, non-invasive, new tool for assessing the health status of this endangered species. This versatile and non-lethal monitoring tool might increase the usefulness of sea turtles as sentinel species, as indicators of the environmental health of their marine-coastal habitats in the light of spatial-temporal changes, and environmental degradation [69].

5. Conclusions

This is the first study that was conducted on loggerhead sea turtles, where the response of erythrocytes to MeHg-induced oxidative stress is evaluated. Erythrocytes resulted in being an excellent, simple, and inexpensive model to perform bioassays for ecotoxicological evaluation with heavy metals and other environmental pollutant molecules, specifically for MeHg in this study. Being loggerhead turtles considered a threatened or endangered species, RBCs represent an excellent ethical experimental in vitro model, allowing ecotoxicological studies to be carried out without the need to sacrifice animals. In addition, SOD, GST, and MDA were sensitive biomarkers of environmental pollution and oxidative stress in loggerhead turtles. However, these biomarkers presented high standard deviations due to the low number of individuals (n = 5) used. Future studies are suggested for carrying out further evaluations with a larger sample to improve the statistical significance of these results and confirm the benefits of these biomarkers as indicators of pollution and oxidative stress. It was observed that exposure to methylmercury in low concentrations induce protective redox defenses in RBCs, as seen from the induction of anti-oxidant enzyme SOD and GST activity. However, when the methylmercury concentration increases, these defenses are overcome and lipid peroxidation injury increases. Erythrocytes of loggerhead turtles in this study survived when kept in plasma in vacutainer tubes with Na-heparin at 4 °C for about eight weeks. This important characteristic allows for the use of these cells in sequential experiments.

Author Contributions: Conceptualization, J.H.-F. and E.A.L.-B.; methodology, J.H.-F., P.R.-B. and E.A.L.-B.; software, J.H.-F. and P.R.-B.; validation, J.H.-F., P.R.-B., L.M.-R., A.P.-V. and E.A.L.-B.; formal analysis, J.H.-F. and E.A.L.-B.; investigation, J.H.-F. and P.R.-B.; resources, J.H.-F., P.R.-B. and E.A.L.-B.; data curation, J.H.-F., P.R.-B., L.M.-R., A.P.-V. and E.A.L.-B.; writing—original draft preparation, J.H.-F. writing—review and editing, J.H.-F., P.R.-B., L.M.-R., A.P.-V. and E.A.L.-B.; supervision, E.A.L.-B., L.M.-R. and A.P.-V.; project administration, J.H.-F.; funding acquisition, J.H.-F., L.M.-R. and E.A.L.-B. All authors have read and agreed to the published version of the manuscript.

Funding: This work was supported by the Office of Research, Creation and Innovation of the Universidad Jorge Tadeo Lozano (340-07-10). Additional funding came from the Intramural Research Program of the National Institutes of Health (NIH, USA), National Library of Medicine (USA), National Center for Biotechnology Information (NCBI) ZIA LM082713–06. There was no additional external funding received for this study.

Acknowledgments: We are grateful to Rodadero Aquarium in Santa Marta for their collaboration in obtaining and providing samples of loggerhead for the development of this study. Samples were obtained under a research permit that was granted by the Ministry of Environment and Territorial Development (#24 of 22 June 2012) Contract for Access to Genetic Resources (#64 of 23 April 2013), and permit for the collection of samples of Colombian biodiversity, issued to the Jorge Tadeo Lozano University (Resolution 1271 of 23 October 2014, IDB040I File). The authors thank Michel Arhens for critical review of the manuscript.

Conflicts of Interest: The authors declare that they have no known competing financial interests or personal relationships that could have influenced the results reported in this paper.

References

1. Finlayson, K.A.; Leusch, F.D.L.; Van De Merwe, J.P. The current state and future directions of marine turtle toxicology research. *Environ. Int.* **2016**, *94*, 113–123. [CrossRef]
2. Wallace, B.P.; Dimatteo, A.D.; Hurley, B.J.; Finkbeiner, E.M.; Alan, B.; Chaloupka, M.Y.; Hutchinson, B.J.; Abreu-grobois, F.A.; Marcovaldi, M.A.; Mortimer, J.A.; et al. Regional Management Units for Marine Turtles: A Novel Framework for Prioritizing Conservation and Research across Multiple Scales. *PLoS ONE* **2010**, *5*. [CrossRef] [PubMed]
3. Lancheros-piliego, D.; Fernández, J.H. AMDAR y PCR-extra-rápida para la identificación de la tortuga cabezona *Caretta caretta* (Testudines: Cheloniidae) utilizando el gen mitocondrial citocromo c oxidasa I (COI). *Univ. Sci.* **2013**, *18*. [CrossRef]
4. Day, R.D.; Segars, A.L.; Arendt, M.D.; Lee, A.M.; Peden-Adams, M.M. Relationship of blood mercury levels to health parameters in the loggerhead sea turtle (*Caretta caretta*). *Environ. Health Perspect.* **2007**. [CrossRef] [PubMed]
5. Amorocho, D. Monitoring nesting loggerhead turtles (*Caretta caretta*) in the central Caribbean coast of Colombia. *Mar. Turt. Newsl.* **2003**, *101*, 8–13.
6. Jule, K.R.; Leaver, L.A.; Lea, S.E.G. The effects of captive experience on reintroduction survival in carnivores: A review and analysis. *Biol. Conserv.* **2008**, *141*, 355–363. [CrossRef]
7. Beyer, W.N.; Meador, J.P. *Environmental Contaminants in Wildlife: Interpreting Tissue Concentrations*; CRC Press: Boca Raton, FL, USA, 1996; pp. 1–768. [CrossRef]
8. Ley-quiñónez, C.; Zavala-norzagaray, A.A.; Espinosa-carreón, T.L.; Peckham, H. Baseline heavy metals and metalloid values in blood of loggerhead turtles (*Caretta caretta*) from Baja California Sur, Mexico. *Mar. Pollut. Bull.* **2011**, *62*, 1979–1983. [CrossRef]
9. Yarsan, E.; Yipel, M. The Important Terms of Marine Pollution "Biomarkers and Biomonitoring. *J. Mol. Biomark. Diagn.* **2013**, 1–4. [CrossRef]
10. Yipel, M.; Tekeli, İ.O.; İşler, C.T.; Altuğ, M.E. Heavy metal distribution in blood, liver and kidneys of Loggerhead (*Caretta caretta*) and Green (*Chelonia mydas*) sea turtles from the Northeast Mediterranean Sea. *Mar. Pollut. Bull.* **2017**, *125*, 487–491. [CrossRef]
11. Yang, L.; Zhang, Y.; Wang, F.; Luo, Z.; Guo, S.; Strähle, U. Toxicity of mercury: Molecular evidence. *Chemosphere* **2020**, *245*. [CrossRef]
12. Kotnik, J.; Horvat, M.; Begu, E.; Shlyapnikov, Y.; Sprovieri, F.; Pirrone, N. Dissolved gaseous mercury (DGM) in the Mediterranean Sea: Spatial and temporal trends. *Mar. Chem.* **2017**, *193*, 8–19. [CrossRef]
13. Burger, J.; Gochfeld, M. Mercury in canned tuna: White versus light and temporal variation. *Environ. Res.* **2004**, *96*, 239–249. [CrossRef] [PubMed]
14. Parra, J.P.; Betancourt, J.; Espinosa, L.F.; Garay, J. Evolución y estado de la contaminación por metales pesados y compuestos orgánicos en la Bahía de Cartagena, Colombia. Informe Técnico. *Inf. Técnico* **2011**, 1–13.
15. Tosic, M.; Restrepo, J.D.; Lonin, S.; Izquierdo, A.; Martins, F. Water and sediment quality in Cartagena Bay, Colombia: Seasonal variability and potential impacts of pollution. *Estuar. Coast. Shelf Sci.* **2019**, *216*, 187–203. [CrossRef]
16. Selin, N.E.; Jacob, D.J.; Yantosca, R.M.; Strode, S.; Jaeglé, L.; Sunderland, E.M. Global 3-D land-ocean-atmosphere model for mercury: Present-day versus preindustrial cycles and anthropogenic enrichment factors for deposition. *Glob. Biogeochem. Cycles* **2008**, *22*, 1–13. [CrossRef]
17. Pirrone, N.; Cinnirella, S.; Feng, X.; Finkelman, R.B.; Friedli, H.R.; Leaner, J.; Mason, R.; Mukherjee, A.B.; Stracher, G.B.; Streets, D.G.; et al. Global mercury emissions to the atmosphere from anthropogenic and natural sources. *Atmos. Chem. Phys.* **2010**, *10*, 5951–5964. [CrossRef]
18. Farina, M.; Aschner, M.; Rocha, J.B.T. Oxidative stress in MeHg-induced neurotoxicity. *Toxicol. Appl. Pharmacol.* **2011**, *256*, 405–417. [CrossRef]
19. Joshi, D.; Kumar, M.D.; Kumar, A.; Sangeeta, S. Reversal of Methylmercury-Induced Oxidative Stress, Lipid Peroxidation, and DNA Damage by the Treatment of N-Acetyl Cysteine: A Protective Approach. *J. Environ. Pathol. Toxicol. Oncol.* **2014**, *33*, 167–182. [CrossRef]
20. Rodríguez, M.; Osuna, A. Efecto de la adición de antioxidantes sobre la motilidad espermática post-criopreservación y fertilidad del semen de peces. *Rev. Vet.* **2017**, *28*, 157–164.

21. Franco, R.; Navarro, G.; Mart, E. Antioxidant defense mechanisms in erythrocytes and in the central nervous system. *Antioxidants* **2019**, *8*, 46. [CrossRef]
22. Orson, R.; Brncunn, L. Uptake of methyl mercuric chloride and mercuric chloride by trout: A sturly of uptake pathways into the whotre anirnal and uptake by erythrocytes in vitro. *J. Fish. Board Can.* **1973**, *30*, 1293–1299. [CrossRef]
23. Pagano, M.; Faggio, C. The use of erythrocyte fragility to assess xenobiotic cytotoxicity. *Cell Biochem. Funct.* **2015**, *33*, 351–355. [CrossRef] [PubMed]
24. Massarsky, A.; Abraham, R.; Nguyen, K.C.; Rippstein, P.; Tayabali, A.F.; Trudeau, V.L.; Moon, T.W. Nanosilver cytotoxicity in rainbow trout (*Oncorhynchus mykiss*) erythrocytes and hepatocytes. *Comp. Biochem. Physiol. Part C Toxicol. Pharmacol.* **2014**, *159*, 10–21. [CrossRef] [PubMed]
25. Gabryelak, T.; Piatkowska, M.; Leyko, W.P.G. Seasonal variations in the activities of peroxide metabolism enzymes in erythrocytes of freshwater fish species. *Comp. Biochem. Physiol. Part C Toxicol. Pharmacol.* **1983**, *75*, 383–385. [CrossRef]
26. Goodchild, C.G.; DuRant, S.E. Fluorescent heme degradation products are biomarkers of oxidative stress and linked to impaired membrane integrity in avian red blood cells. *Physiol. Biochem. Zool.* **2020**, *93*, 129–139. [CrossRef]
27. Velando, A.; Noguera, J.C.; da Silva, A.; Kim, S.Y. Redox-regulation and life-history trade-offs: Scavenging mitochondrial ROS improves growth in a wild bird. *Sci. Rep.* **2019**, *9*, 1–9. [CrossRef]
28. Sakuragui, M.M.; Paulino, M.G.; da Silva e Souza, N.E.; Tavares, D.; Terezan, A.P.; Pesenti, E.; Giani, A.; Fernandes, J.B.; Cestari, M.M.; Fernandes, M.N. Crude extract of cyanobacterium Radiocystis fernandoi strain R28 induces anemia and oxidative stress in fish erythrocytes. *Toxicon* **2019**, *169*, 18–24. [CrossRef]
29. Cortés-Gutiérrez, E.I.; García-Salas, J.A.; Dávila-Rodríguez, M.I.; Ceyca-Contreras, J.P.; González-Ramírez, E.G. Evaluation of oxidative DNA damage in pigeon erythrocytes using DNA breakage detection-fluorescence in situ hybridization (DBD-FISH). *Biotech. Histochem.* **2019**, *94*, 600–605. [CrossRef]
30. Burgos-Aceves, M.A.; Lionetti, L.; Faggio, C. Multidisciplinary haematology as prognostic device in environmental and xenobiotic stress-induced response in fish. *Sci. Total Environ.* **2019**, *670*, 1170–1183. [CrossRef]
31. Panghal, A.; Sathua, K.B.; Flora, S.J.S. Gallic acid and MiADMSA reversed arsenic induced oxidative/nitrosative damage in rat red blood cells. *Heliyon* **2020**, *6*, e03431. [CrossRef]
32. Hammami, N.; Athmouni, K.; Lahmar, I.; Ben Abdallah, F.; Belghith, K. Antioxidant potential of Salicornia arabica lipid extract and their protective effect against cadmium induced oxidative stress in erythrocytes isolated from rats. *J. Food Meas. Charact.* **2019**, *13*, 2705–2712. [CrossRef]
33. Agrawal, D.; Sultana, P.; Gupta, G.S.D. Oxidative damage and changes in the glutathione redox system in erythrocytes from rats treated with hexachlorocyclohexane. *Food Chem. Toxicol.* **1991**, *29*, 459–462. [CrossRef]
34. Barraza, A.D.; Komoroske, L.M.; Allen, C.; Eguchi, T.; Gossett, R.; Holland, E.; Lawson, D.D.; LeRoux, R.A.; Long, A.; Seminoff, J.A.; et al. Trace metals in green sea turtles (*Chelonia mydas*) inhabiting two southern California coastal estuaries. *Chemosphere* **2019**, *223*, 342–350. [CrossRef] [PubMed]
35. Çördük, N.; Doğru, N.H.; Gül, Ç.; Tosunoğlu, M. Assessment of nuclear abnormalities in erythrocytes of balkan pond turtle *Mauremys rivulata* (Valenciennes, 1833) (Testudines: Geoemydidae) from the Biga Stream, Çanakkale, Turkey. *Acta Zool. Bulg.* **2019**, *71*, 219–226.
36. Aguirre, A.A.; Balazs, G.H. Comparative Original Article Blood Biochemistry Values of Green Turtles, *Chelonia Mydas*, With and Without Fibropapillomatosis. *Comp. Haematol. Int.* **2000**, *10*, 132–137. [CrossRef]
37. Bjorndal, K.A.; Bolten, A.B.; Martins, H.R. Somatic growth model of juvenile loggerhead sea turtles *Caretta caretta*: Duration of pelagic stage. *Mar. Ecol. Prog. Ser.* **2000**, *202*, 265–272. [CrossRef]
38. Dutton, P.H. Methods for collection and preservation of samples for sea turtle genetic studies. In Proceedings of the International Symposium on Sea Turtle Conservation Genetics, Miami, FL, USA, 1 January 1996.
39. Finney, D.T. *Probit Analysis*; Cambridge University Press: Cambridge, UK, 1971.
40. Venables, W.N.; Ripley, B.D. *Modern Applied Statistics with S*; Springer Science & Business Media: New York, NY, USA, 2002.
41. Guisande, C.; Heine, J.; González-DaCosta, J.; García-Roselló, E. *RWizard Software*; University of Vigo: Vigo, Spain, 2014.

42. Martins, R.D.P.; Braga, H.D.C.; Aline, P.; Dalmarco, J.B.; De Bem, A.F.; Roberto, A.; Santos, S.; Dafre, A.L.; Pizzolatti, M.G.; Latini, A.; et al. Synergistic neurotoxicity induced by methylmercury and quercetin in mice. *Food Chem. Toxicol.* **2009**, *47*, 645–649. [CrossRef]
43. Habig, W.H.; Pabst, M.J.; Jakoby, W.B. Glutathione S-Transferases The First Enzymatic Step In Mercapturic Acid Formation. *J. Biol. Chem.* **2005**, *249*, 7130–7139.
44. R Development Core Team, R. *A Language and Environment for Statistical Computing*; R Foundation for Statistical Computing: Copenhagen, Denmark, 2017; p. 900051.
45. Casal, A.B.; Luis, F.L.; Juste, C.; Or, J. Comparative study of hematologic and plasma biochemical variables in Eastern Atlantic juvenile and adult nesting loggerhead sea turtles (*Caretta caretta*). *Vet. Clin. Pathol.* **2009**, *2*, 213–218. [CrossRef]
46. Deem, S.L.; Norton, T.M.; Mitchell, M.; Segars, A.; Alleman, A.R.; Cray, C.; Poppenga, R.H.; Dodd, M.; Karesh, W.B. Comparison of blood values in foraging, nesting, and stranded loggerhead turtles (*Caretta caretta*) along the coast of Georgia, USA. *J. Wildl. Dis.* **2009**, *45*, 41–56. [CrossRef]
47. Yang, T.; Haas, H.L.; Patel, S.; Smolowitz, R.; James, M.C.; Williard, A.S. Blood biochemistry and haematology of migrating loggerhead turtles (Caretta caretta) in the Northwest Atlantic: Reference intervals and intra-population comparisons. *Conserv. Physiol.* **2019**, *7*, 1–15. [CrossRef] [PubMed]
48. Rousselet, E.; Stacy, N.I.; LaVictoire, K.; Higgins, B.M.; Tocidlowski, M.E.; Flanagan, J.P.; Godard-Codding, C.A. Hematology and plasma biochemistry analytes in five age groups of immature, captive-reared loggerhead sea turtles (Caretta caretta). *J. Zoo Wildl. Med.* **2013**, *44*, 859–874. [CrossRef] [PubMed]
49. Angeles, M.; Morcillo, P.; Cuesta, A. In vitro effects of metals on isolated head-kidney and blood leucocytes of the teleost fi sh *Sparus aurata* L. and *Dicentrarchus labrax* L. *Fish Shellfish Immunol.* **2016**, *54*. [CrossRef]
50. Lutz, P.L.; Dunbar-Cooper, A.N.N. The loggerhead sea turtle, Caretta Caretta. *Fish. Bull.* **1971**, *85*, 37.
51. Kelly, T.R.; McNeill, J.B.; Avens, L.; Hall, A.G.; Goshe, L.R.; Hohn, A.A.; Godfrey, M.H.; Nicole Mihnovets, A.; Cluse, W.M.; Harms, C.A. Clinical pathology reference intervals for an in-water population of juvenile loggerhead sea turtle s (*Caretta caretta*) in Core Sound, North Carolina, USA. *PLoS ONE* **2015**, *10*, 1–13. [CrossRef] [PubMed]
52. Flint, M.; Morton, J.M.; Limpus, C.J.; Patterson-Kane, J.C.; Mills, P.C. Reference intervals for plasma biochemical and hematologic measures in loggerhead sea turtles (Caretta caretta) from Moreton bay Australia. *J. Wildl. Dis.* **2010**, *46*, 731–741. [CrossRef] [PubMed]
53. Ruggiero, M.G.; Ferretti, L.; Glomski, C.; Federico, N. Erythrophagocytosis in Circulating Blood of Loggerhead Turtles *Caretta caretta*: The Pitting of Heinz Bodies. *J. Exp. Zool.* **2013**. [CrossRef]
54. Figueres, J. *Estudio Sanitario de las Tortugas Terrestres Mediterráneas (Género Testudo) e Implicaciones Para su Conservación*; Universidad Autonoma de Barcelona: Barcelona, Spain, 2015; pp. 1–187.
55. Finlayson, K.A.; Leusch, F.D.L.; Van De Merwe, J.P. Science of the Total Environment Cytotoxicity of organic and inorganic compounds to primary cell cultures established from internal tissues of *Chelonia mydas*. *Sci. Total Environ.* **2019**, *664*, 958–967. [CrossRef]
56. Olivero-verbel, J.; Caballero-gallardo, K.; Torres, N. International Journal of Environmental Assessment of mercury in muscle of fish from Cartagena Bay, a tropical estuary at the north of Colombia. *Int. J. Environ. Health Res.* **2009**, *5*, 37–41. [CrossRef]
57. Stricks, W.; Kolthoff, I.M. Reactions between Mercuric Mercury and Cysteine and Glutathione. Apparent Dissociation Constants, Heats and Entropies of Formation of Various Forms of Mercuric Mercapto-Cysteine and -Glutathione. *J. Am. Chem. Soc.* **1953**, *75*, 5673–5681. [CrossRef]
58. Weed, R.; Eber, J.; Biology, R. Interaction of Mercury with Human Erythrocytes. *J. Gen. Physiol.* **1962**, *45*, 395–410. [CrossRef] [PubMed]
59. Naganuma, A.; Koyama, Y.; Imura, N. Behavior of Methylmercury in Mammalian Erythrocytes. *Toxicol. Appl. Pharmacol.* **1980**, *410*, 405–410. [CrossRef]
60. Farina, M.; Aschner, M. Methylmercury-Induced Neurotoxicity: Focus on Pro-oxidative Events and Related Consequences. In *Neurotoxicity of Metals*; Aschner, M., Costa, L.G., Eds.; Springer International Publishing: Cham, Switzerland, 2017; pp. 267–286, ISBN 978-3-319-60189-2.
61. Zapata, L.M.; Bock, B.C.; Yaneth, L.; Palacio, J.A. Ecotoxicology and environmental fafety application of the micronucleus test and comet assay in *Trachemys callirostris* erythrocytes as a model for in situ genotoxic monitoring. *Ecotoxicol. Environ. Saf.* **2016**, *127*, 108–116. [CrossRef] [PubMed]

62. Cocci, P.; Capriotti, M.; Mosconi, G.; Palermo, F.A. Effects of endocrine disrupting chemicals on estrogen receptor alpha and heat shock protein 60 gene expression in primary cultures of loggerhead sea turtle (*Caretta caretta*) erythrocytes ☆. *Environ. Res.* **2017**, *158*, 616–624. [CrossRef] [PubMed]
63. Tan, F.; Wang, M.; Wang, W.; Alonso Aguirre, A.; Lu, Y. Validation of an in vitro cytotoxicity test for four heavy metals using cell lines derived from a green sea turtle (*Chelonia mydas*). *Cell Biol. Toxicol.* **2010**, *26*, 255–263. [CrossRef] [PubMed]
64. Tong, J.; Wang, Y.; Lu, Y. In vitro evaluation of inorganic and methyl mercury mediated cytotoxic effect on neural cells derived from different animal species. *J. Environ. Sci.* **2016**, *41*, 138–145. [CrossRef]
65. Young, J.L.; Wise, S.S.; Xie, H.; Zhu, C.; Fukuda, T.; Wise, J.P. Comparative cytotoxicity and genotoxicity of soluble and particulate hexavalent chromium in human and hawksbill sea turtle (*Eretmochelys imbricata*) skin cells. *Comp. Biochem. Physiol. Part C Toxicol. Pharmacol.* **2015**, *178*, 145–155. [CrossRef]
66. Kaska, Y.; Furness, R.W. Heavy metals in marine turtle eggs and hatchlings in the mediterranean. *Zool. Middle East* **2001**, *24*, 127–132. [CrossRef]
67. Kampalath, R.; Gardner, S.C.; Méndez-Rodríguez, L.; Jay, J.A. Total and methylmercury in three species of sea turtles of Baja California Sur. *Mar. Pollut. Bull.* **2006**, *52*, 1816–1823. [CrossRef]
68. Mcgraw, K.J.; Cohen, A.A.; Costantini, D.; Ho, P. The ecological significance of antioxidants and oxidative stress: A marriage between mechanistic and functional perspectives. *Funct. Ecol.* **2010**, *2000*, 947–949. [CrossRef]
69. Labrada-martagón, V.; Rodríguez, P.A.T.; Méndez-rodríguez, L.C.; Zenteno-savín, T. Comparative Biochemistry and Physiology, Part C Oxidative stress indicators and chemical contaminants in East Paci fi c green turtles (*Chelonia mydas*) inhabiting two foraging coastal lagoons in the Baja California peninsula. *Comp. Biochem. Physiol. Part C: Toxicol. Pharmacol.* **2011**, *154*, 65–75. [CrossRef]
70. Perrault, J.R.; Schmid, J.R.; Walsh, C.J.; Yordy, J.E.; Tucker, A.D. Brevetoxin exposure, superoxide dismutase activity and plasma protein electrophoretic profiles in wild-caught Kemp's ridley sea turtles (*Lepidochelys kempii*) in southwest Florida. *Harmful Algae* **2014**, *37*, 194–202. [CrossRef]
71. Finlayson, K.A.; Leusch, F.D.L.; Van De Merwe, J.P. Primary green turtle (*Chelonia mydas*) skin fibroblasts as an in vitro model for assessing genotoxicity and oxidative stress. *Aquat. Toxicol.* **2018**. [CrossRef] [PubMed]
72. Isaji, Y.; Okochi, M.; Horio, F.; Honda, H. Use of erythrocyte adhesion assay to predict the risk of diabetic complications. *Biochem. Eng. J.* **2009**, *43*, 178–184. [CrossRef]
73. Schara, M.; Nemec, M.; Falnoga, I.; Kobal, A.B.; Kveder, M.; Svetek, J. The action of mercury on cell membranes. *Cell. Mol. Biol. Lett.* **2001**, *6*, 299–304. [PubMed]
74. Cortés-Gómez, A.A.; Romero, D.; Girondot, M. The current situation of inorganic elements in marine turtles: A general review and meta-analysis. *Environ. Pollut.* **2017**, *229*, 567–585. [CrossRef]
75. Salvarani, P.I.; Vieira, L.R.; Ku-Peralta, W.; Morgado, F.; Osten, J.R. Von Oxidative stress biomarkers and organochlorine pesticides in nesting female hawksbill turtles *Eretmochelys imbricata* from Mexican coast (Punta Xen, Mexico). *Environ. Sci. Pollut. Res.* **2018**, *25*, 23809–23816. [CrossRef]
76. Berntssen, M.H.G.; Aatland, A.; Handy, R.D. Chronic dietary mercury exposure causes oxidative stress, brain lesions, and altered behaviour in Atlantic salmon (*Salmo salar*) parr. *Aquat. Toxicol.* **2003**, *65*. [CrossRef]
77. Zaman, K.; Macgill, R.S.; Johnson, J.E.; Ahmad, S.; Pardini, R.S. An Insect Model for Assessing Mercury Toxicity: Effect of Mercury on Antioxidant Enzyme Activities of the Housefly (*Musca domestica*) and the Cabbage Looper Moth (*Trichoplusia ni*). *Arch. Environ. Contam. Toxicol.* **1994**, *118*, 114–118. [CrossRef]
78. Hirota, Y.; Yamaguchi, S.; Shimojoh, N.; Sano, K.I. Inhibitory effect of methylmercury on the activity of glutathione peroxidase. *Toxicol. Appl. Pharmacol.* **1980**, *176*, 174–176. [CrossRef]
79. Kumagai, Y.; Shinyashiki, M.; Shimojo, N. Alterations in Superoxide Dismutase Isozymes by Methylmercury. *Appl. Organomet. Chem.* **1997**, *11*, 635–643. [CrossRef]
80. Lund, B.; Miller, M.; Woodst, S. Studies on Hg (II) -Induced H_2O_2 formation and oxidative stress in vivo and in vitro in rat. *Biochmdcal Phamwcology.* **2017**, *45*, 2017–2024.
81. Yee, S.; Choi, B. Methylmercury poisoning induces oxidative stress in the mouse brain. *Exp. Mol. Pathol.* **1994**, *60*, 188–196. [CrossRef] [PubMed]
82. Clarkson, T.W. The Toxicology of Mercury and Its Chemical Compounds. *Crit. Rev. Toxicol.* **2006**, *36*, 609–662. [CrossRef] [PubMed]
83. Stier, A.; Reichert, S.; Criscuolo, F.; Bize, P. Red blood cells open promising avenues for longitudinal studies of ageing in laboratory, non-model and wild animals. *Exp. Gerontol.* **2015**, *71*, 118–134. [CrossRef] [PubMed]

84. Tanaka, R.; Nakai, K.; Nakai, K. Hemolysis and morphological changes in rat erythrocytes with mercurials. *Jpn. J. Pharmacol.* **1977**, *27*, 413–419. [CrossRef]
85. Lillig, C.H.; Holmgren, A. Thioredoxin and Related Molecules–From Biology to Health and Disease. *Antioxid. Redox Signalin* **2006**, *9*, 25–47. [CrossRef]
86. Hall, A.; Karplus, P.A.; Poole, L.B. Typical 2-Cys peroxiredoxins - Structures, mechanisms and functions. *FEBS J.* **2009**, *276*, 2469–2477. [CrossRef]
87. Perkins, A.; Nelson, K.J.; Parsonage, D.; Poole, L.B.; Karplus, P.A. Peroxiredoxins: Guardians against oxidative stress and modulators of peroxide signaling. *Trends Biochem. Sci.* **2015**, *40*, 435–445. [CrossRef]
88. Arnér, E.S.J. Focus on mammalian thioredoxin reductases—Important selenoproteins with versatile functions. *Biochim. Biophys. Acta (BBA)-Gen. Subj.* **2009**, *1790*, 495–526. [CrossRef]
89. Winterbourn, C.C. *The Biological Chemistry of Hydrogen Peroxide*, 1st ed.; Elsevier Inc.: Amsterdam, The Netherlands, 2013; Volume 528, ISBN 9780124058811.
90. Zhang, Y.; Mi, K.; Ding, X.; Li, Y.; Wang, T.; Dou, T.; Ding, J.; Wei, W. Characterization of a classical 2-cysteine peroxiredoxin1 gene from Chinese soft-shelled turtle *Pelodiscus sinensis* with its potent antioxidant properties and putative immune response. *Dev. Comp. Immunol.* **2019**. [CrossRef] [PubMed]
91. Carvalho, C.M.L.; Chew, E.; Hashemy, S.I.; Lu, J. Inhibition of the Human Thioredoxin System. *J. Biol. Chem.* **2008**, *283*, 11913–11923. [CrossRef] [PubMed]
92. Farina, M.; Avila, D.S.; Da Rocha, J.B.T.; Aschner, M. Metals, oxidative stress and neurodegeneration: A focus on iron, manganese and mercury. *Neurochem. Int.* **2013**, *62*, 575–594. [CrossRef] [PubMed]
93. Maniero, M.Á.; Wuilloud, R.G.; Callegari, E.A.; Smichowski, P.N.; Fanelli, M.A. Metalloproteomics analysis in human mammary cell lines treated with inorganic mercury. *J. Trace Elem. Med. Biol.* **2020**, *58*, 126441. [CrossRef]
94. Liu, D.L.; Zhao, L.X.; Zhang, S.; Du, J.R. Peroxiredoxin 1-mediated activation of TLR4/NF-κB pathway contributes to neuroinflammatory injury in intracerebral hemorrhage. *Int. Immunopharmacol.* **2016**, *41*, 82–89. [CrossRef] [PubMed]
95. Poynton, R.A.; Hampton, M.B. Peroxiredoxins as biomarkers of oxidative stress. *Biochim. Biophys. Acta (BBA)-Gen. Subj.* **2014**, *1840*, 906–912. [CrossRef]
96. Immenschuh, S.; Baumgart-Vogt, E. Peroxiredoxins, oxidative stress, and cell proliferation. *Antioxid. Redox Signal.* **2005**, *7*, 768–777. [CrossRef]
97. Walsh, C.J.; Cocilova, C.; Restivo, J.; Flewelling, L.; Milton, S. Immune function in Trachemys scripta following exposure to a predominant brevetoxin congener, PbTx-3, as a model for potential health impacts for sea turtles naturally exposed to brevetoxins. *Ecotoxicol* **2019**, *28*, 1085–1104. [CrossRef]
98. Krivoruchko, A.; Storey, K.B. Activation of antioxidant defenses in response to freezing in freeze-tolerant painted turtle hatchlings. *Biochim. Biophys. Acta (BBA)-Gen. Subj.* **2010**. [CrossRef]
99. Storey, K.B. Reptile freeze tolerance: Metabolism and gene expression. *Cryobiology* **2006**, *52*, 1–16. [CrossRef]
100. Xu, J.; Zhao, J.; Li, Y.; Zou, Y.; Lu, B.; Chen, Y.; Ma, Y.; Xu, H. Evaluation of differentially expressed immune-related genes in intestine of Pelodiscus sinensis after intragastric challenge with lipopolysaccharide based on transcriptome analysis. *Fish Shellfish Immunol.* **2016**. [CrossRef] [PubMed]
101. Branco, V.; Ramos, P.; Canário, J.; Lu, J.; Holmgren, A.; Carvalho, C. Biomarkers of adverse response to mercury: Histopathology versus thioredoxin reductase activity. *BioMed Res. Int.* **2012**, *2012*. [CrossRef]
102. Robitaille, S.; Mailloux, R.J.; Man, H. Methylmercury alters glutathione homeostasis by inhibiting glutaredoxin 1 and enhancing glutathione biosynthesis in cultured human astrocytoma cells. *Toxicol. Lett.* **2016**, *256*, 1–10. [CrossRef] [PubMed]
103. Farina, M.; Campos, F.; Vendrell, I.; Berenguer, J.; Barzi, M. Probucol Increases Glutathione Peroxidase-1 Activity and Displays Long-Lasting Protection against Methylmercury Toxicity in Cerebellar Granule Cells. *Toxicol. Sci.* **2009**, *112*, 416–426. [CrossRef] [PubMed]
104. Carvalho, C.M.L.; Lu, J.; Zhang, X.; Arne, E.S.J.; Effects, A. Effects of selenite and chelating agents on mammalian thioredoxin reductase inhibited by mercury: Implications for treatment of mercury poisoning. *FASEB J.* **2010**, *25*, 370–381. [CrossRef]
105. Shinyashiki, M.; Kumagai, Y.; Homma-Takeda, S.; Nagafune, J.; Shimojo, N. Selective inhibition of the mouse brain Mn-SOD by methylmercury. *Environ. Toxicol. Pharmacol.* **1996**, *2*, 359–366. [CrossRef]

106. Toyama, T.; Sumi, D.; Shinkai, Y.; Yasutake, A.; Taguchi, K.; Tong, K.I.; Yamamoto, M.; Kumagai, Y. Cytoprotective role of Nrf2 / Keap1 system in methylmercury toxicity. *Biochem. Biophys. Res. Commun.* **2007**, *363*, 645–650. [CrossRef]
107. Messer, R.L.W.; Lockwood, P.E.; Tseng, W.Y.; Edwards, K.; Shaw, M.; Caughman, G.B.; Lewis, J.B.; Wataha, J.C.; College, M. Mercury (II) alters mitochondrial activity of monocytes at sublethal doses via oxidative stress mechanisms. *J. Biomed. Mater. Res. Part B* **2005**, 257–263. [CrossRef]
108. Wataha, J.C.; Lewis, J.B.; Mccloud, V.V.; Shaw, M.; Omata, Y.; Lockwood, P.E.; Messer, R.L.W.; Hansen, J.M. Effect of mercury (II) on Nrf2, thioredoxin reductase-1 and thioredoxin-1 in human monocytes. *Dent. Mater.* **2007**, *24*, 765–772. [CrossRef]

applied sciences

MDPI

Article

Total Mercury in Soils and Sediments in the Vicinity of Abandoned Mercury Mine Area in Puerto Princesa City, Philippines

Jessie Samaniego *, Cris Reven Gibaga, Alexandria Tanciongco and Rasty Rastrullo

Department of Science and Technology-Philippine Nuclear Research Institute, Commonwealth Ave., Diliman, Quezon City 1101, Philippines; crlgibaga@pnri.dost.gov.ph (C.R.G.); amtanciongco@up.edu.ph (A.T.); rmrastrullo@gmail.com (R.R.)

* Correspondence: josamaniego@pnri.dost.gov.ph

Received: 28 October 2019; Accepted: 18 December 2019; Published: 2 July 2020

Abstract: The abandoned mercury (Hg) mine area in Puerto Princesa City, Palawan was included in the list of abandoned and inactive mines in the Philippines which pose a high risk to human health and the environment, and require rehabilitation. The mine site, operated by Palawan Quicksilver Mines, Inc. (PQMI) from 1953 to 1976, is located approximately 3 km inland from Honda Bay coast and within the catchment of the Tagburos River, which is a local fishery and recreational area. In this study, total Hg levels in soils and sediments were measured to assess the possible release of Hg from the site into the surrounding natural environment. Results showed that total Hg (THg) concentrations found in soils (0.04–67.5 mg kg^{-1}), mine waste calcines (52.7–924.2 mg kg^{-1}), river sediments (1.8–119 mg kg^{-1}), and marine sediments (0.04–12.7 mg kg^{-1}) were elevated compared to the global background of 0.045–0.16 mg kg^{-1}. The high concentrations of Hg in soils and river sediments were influenced by the different pathways for the release of Hg from its mine operations; while Hg in marine sediments was caused by the erosion of mine waste calcines near the pit lake, and calcines used to construct a wharf at the nearby Honda Bay. Mine wastes represent the largest source of Hg contamination in the area, due to the low efficiency of the recovery process during calcination. This work corresponds with the Minamata Convention on Mercury on the updated environmental assessments of abandoned Hg mines as potential source sites of mercury contamination.

Keywords: mercury; PQMI; Palawan; abandoned mines; mine wastes; sediments

1. Introduction

Mining has been an important economic activity in the Philippines, which started centuries ago. With a lack of environmental regulations until recent times, old mining operations have led to many abandoned mine sites, and consequently resulted in high heavy metal concentrations in soils and sediments in the vicinity. In the Philippines, there are more than 20 mine sites that are classified as inactive and abandoned, one of which is the Palawan Quicksilver Mines, Inc. (PQMI) in Puerto Princesa City, Palawan [1]. This former mine site is considered as one of the Hg waste hot spots of the world, because the Hg pollution is affecting the marine environment and the people living nearby [2].

Most of the mercury mining areas around the world have been investigated in relation to their Hg pollution in their local ecosystems, which pose potential risks to humans living nearby. In the past decades, risk assessments on soils and sediments contaminated with Hg from the mining activities have been studied in Hg mining sites in Spain [3,4], China [5,6], New Zealand [7], United States of America [8], and others. Similar patterns of high to extremely high (46,000 mg kg^{-1}) Hg concentrations in mine waste calcines and runoff sediments were reported from these Hg mining sites, which pose potential contamination on soil, water, and air in the area [7].

In the Philippines, a naturally occurring deposit of Hg can be found only in Puerto Princesa City on the island of Palawan. It mostly occurs in cinnabar (HgS) ores hosted by opalite bodies deposited along fractures and weak zones in the bedrock [9,10]. From 1953 to 1976, PQMI mined the Hg deposit in the village areas of Sta. Lourdes and produced around 2900 metric tons of mercury [11]. Today, the abandoned open-pit mine is filled with water, and transformed into a pit lake located about 3 km west from Honda Bay coast (Figure 1). During the mining operations of PQMI, around 2,000,000 metric tons of cinnabar mine waste calcines were produced [11]. Part of those mine waste calcines were used to construct a 600 m by 50 m and 3 m thick wharf at the nearby Honda Bay to serve as port for mining operations from 1953 to 1976. The wharf is now used as port for tourist and fishing boats. The rest of the mine waste calcines were stocked northeast near the pit lake shore (Figure 2), which was bare and had little vegetation until now. For the past four decades, remediation work has not been carried out at this site. Today, the abandoned Hg mine area is in the list of abandoned and inactive mines in the Philippines which pose a high risk to human health and the environment, and which require rehabilitation [1].

Mercury is a highly toxic element that is found both naturally, and as an artificially introduced contaminant, in the environment. Mercury mine sites remain as significant sources of Hg pollution in the environment, due to the unconverted HgS ores left in the mine waste calcines. The calcines contain soluble Hg minerals such as montroydite, eglestonite, corderoite, and elemental Hg. Leaching of these calcines, which has been reported from several abandoned Hg mines, becomes a significant source of Hg if released into the environment [7,11–14]. These anthropogenic activities can create widespread Hg contamination in soils of abandoned mine areas [15], while bare and unvegetated soils can produce dust with particulate-bound Hg during summer, that may be inhaled by the residents. Exposed HgS and mine waste calcines pose a very serious health problem for the residents, as the absorbed Hg from HgS and calcines is mainly accumulated in the kidneys, resembling the disposition pattern of inorganic Hg [16]. Inorganic Hg found in HgS and calcines, may be moved along with the sediments and through water runoff. It can undergo methylation by microorganisms and convert it to methylmercury (MeHg) [17]. Methylmercury may bioaccumulate, and concentrate in the marine and freshwater food chain that might end up in human body [17,18]. Excess concentrations of Hg in the human body can cause serious health threats, like damage to the central nervous system, and is detrimental to fetal development [17–19].

Studies conducted in the abandoned PQMI mine area since 1994 show that the population in the affected areas have relatively high Hg concentrations in their hair and blood samples [20–22]. In 1994, health reports revealed significant elevation of blood Hg levels exceeding the previous recommended exposure level of 20 ppb in 12 out of the 43 residents examined, after their complaints of unusual symptoms (e.g., miscarriages, tooth loss, muscle weakness, paralysis, anemia, tremors, etc.) [21]. Recent reports from the Department of Health (DOH) and Department of Environment and Natural Resources (DENR) say that 33–40% of the 10,000 combined Tagburos and Sta. Lourdes residents have 'chronic mercury poisoning', from the exposure to mine tailings and ingestion of marine products with high Hg content [23]. However, a recent survey of the surface water quality of different water bodies in the vicinity of the PQMI mined area showed undetected Hg concentrations in the pit lake and river, and low Hg concentrations in coastal water near the wharf (0.001 mg L^{-1}) and in hot spring water ($\leq$0.0004 mg L^{-1}) [1]. The very low Hg concentrations measured in the surface water suggest that there is low solubility and bioavailability of inorganic Hg from HgS, and mine waste calcines. For the past 20 years after the last known study on Hg contamination in the area, Hg from HgS and calcines might be carried to the river by erosion and surface runoff, and eventually deposited in Honda Bay where they become a source of contamination and enter the food chain.

The purpose of this study is to document the current status of Hg pollution in different sites surrounding the abandoned PQMI mine area, describe how legacy mining still affects the area, and how it has evolved since the last studies performed in the 1990s. To evaluate the potential Hg contamination in this area, Hg concentrations were measured in the PQMI pit lake, Tagburos River, Honda Bay,

Honda Bay Wharf, mine waste calcines, soils (including agricultural soils) in the vicinity, and other rivers which are not tributaries of the Tagburos River. Assessments on the Hg concentration in fish and the extent of human exposure to Hg were not included in the study. The output of this study will serve as guide to the rehabilitation prioritization of the abandoned PQMI mine area. This study will also conform with the requirements of the Minamata Convention on Mercury, where Philippines is one of the signatories, and give updates on the current environmental conditions of possible point sources of Hg, such as abandoned Hg mining sites.

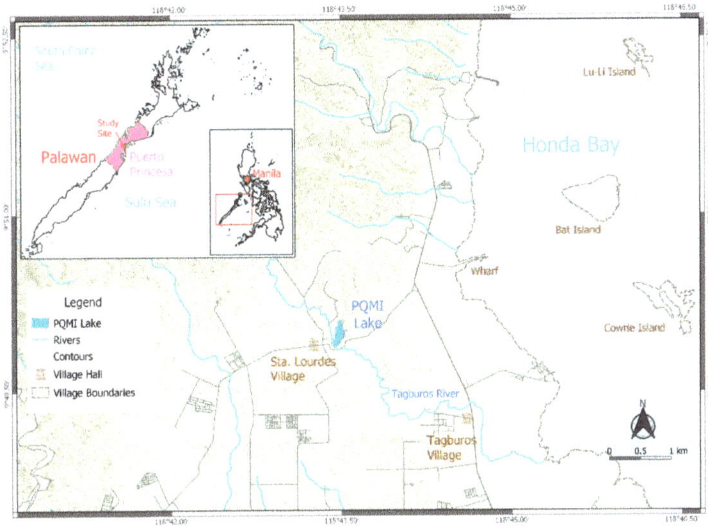

Figure 1. Map of the study site showing the rivers, pit lake, and Honda Bay, with a map of the Philippines and Palawan island (inset).

Figure 2. Photo of mine waste calcines stockpiled at the shore of the pit lake.

2. Materials and Methods

Analyses of total Hg concentrations in different soils and sediments in the vicinity of the abandoned PQMI mine area were conducted to determine the current Hg concentrations in different sub-areas, especially in Tagburos River and Honda Bay. Characterization studies on the Hg contamination in this area is very limited, with the last known published article in 1996 [21]. In 1999, sedimentation rate in Honda Bay was estimated at 1 cm yr^{-1} based on the ^{210}Pb dating of core samples collected from Honda Bay [10]. With a wider scope of sampling area considered in this study, including other rivers and mountainous terrain nearby, assessment of the extent of Hg contamination in the abandoned mine area can be drawn.

2.1. Study Area and Sampling Sites

The study area, Puerto Princesa City, is located in the midsection of Palawan Island. It is bound to the east by the Sulu Sea, and to the west by the South China Sea (Figure 1). The area in Puerto Princesa City with reported human Hg risk encompasses two villages: Sta. Lourdes and Tagburos. The study site, located around 14 km north of Puerto Princesa City center (118°43′ N, 09°50′ E), is approximately 3 km inland from the Honda Bay coast, and situated within the catchment of the Tagburos River, which is a local fishery and tourist recreational area. In this area, a naturally occurring deposit of Hg ore (cinnabar), which is hosted by the Tagburos Opalite formation, can be found. Cinnabar was mined from 1953 to 1976, and the open pit mine site was naturally filled with water, forming a lake.

2.2. Sampling

A total of 102 soil and sediment samples were collected from the study area, between September 2018 to October 2019. Soil samples were collected from the vicinity of former mining sites; while surface sediments were collected from the bottom of the pit lake, Honda Bay, Tagburos River, and other rivers not traversing the pit lake in the area. Mine waste calcine samples were collected from the stockpile near the pit lake and from the scattered calcines situated 500 m northeast of the pit lake. All samples were classified into seven groups: PQMI pit lake, Tagburos River, Honda Bay, Honda Bay Wharf, mine waste calcines, soils (including agricultural soils) in the vicinity, and other rivers which are not tributaries of Tagburos River. Samples were collected using a plastic trowel, from the top 10 cm of the soil surface and surface sediment of rivers. Sediments from Honda Bay and the pit lake were collected using a sediment grab sampler (Wildco Petite Ponar® Grab Sampler, Yulee, FL, USA), collecting the top 5–10 cm of surface sediments. All collected soil and sediment samples were placed in polyethylene resealable plastic and transported to the laboratory at the Philippine Nuclear Research Institute in Manila.

2.3. Total Mercury Analysis

Analyses of THg in the soil and sediment samples were performed at the laboratory of the Philippine Nuclear Research Institute. Samples were prepared by air-drying and sieving up to US Standard 230 mesh (63 μm), to separate the silt-clay (<63 μm) size particles from the sand fraction. Only the <63 μm fractions were analyzed for THg concentrations. Total Hg analyses in the soil and sediment samples were determined using a direct mercury analyzer (Milestone DMA-80 evo, Sorisole, BG, Italy) following the U.S. EPA Method 7473 (mercury in solids and solutions by thermal decomposition, amalgamation, and atomic absorption spectrometry). This Hg analyzer is capable of determining THg concentrations in solid, liquid, and gas samples. Soil samples were weighed (0.10 g) and placed in a nickel boat without any pre-treatment. The boat is then introduced to the direct mercury analyzer where it is initially dried at up to 250 °C, and then thermally decomposed at 900 °C in a continuous flow of oxygen. The THg content is determined using atomic absorption spectrometry at 254 nm. All reported THg concentrations of soil and sediments analyzed were the average of three replicates. The THg analysis results from DMA-80 were validated using certified reference material for stream sediments (GSD-1) prepared by the Institute of Geophysical and Geochemical Exploration of China, with a mean (S.D.) THg value of 0.018 (0.004) mg kg^{-1}. The mean (S.D.) value of the analyzed results using DMA-80 was 0.019 (0.002) mg kg^{-1}, which is in good agreement with the certified value.

3. Results and Discussion

3.1. THg in Mine Waste Calcines and Honda Bay Wharf

From all the surface sediments collected in the vicinity of the abandoned mining site, there are two noticeable hotspots of very high Hg concentration relative to their surroundings (Figure 3). These are the mine waste calcines near the pit lake, and the calcine used in the construction of Honda Bay wharf.

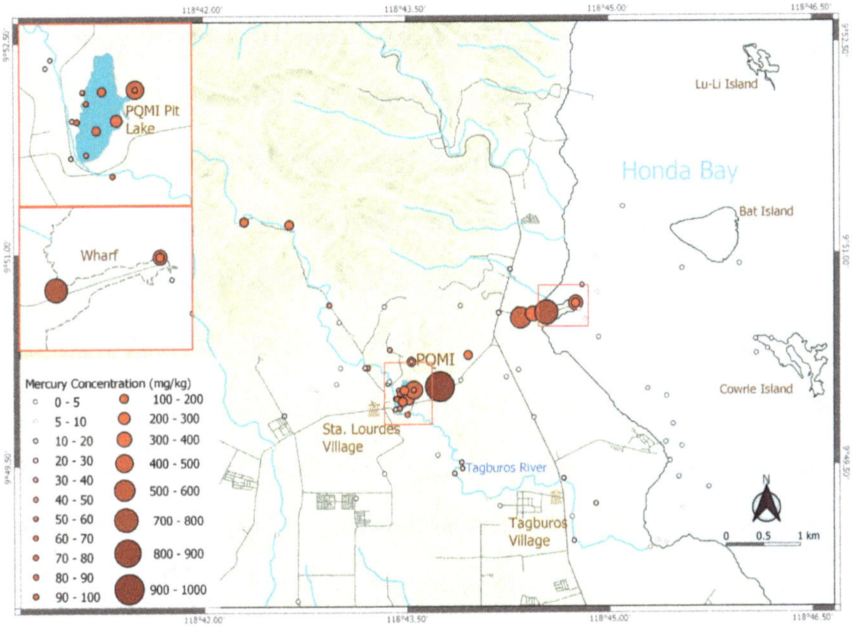

Figure 3. Map showing the total Hg concentration in soils and sediments in the study area.

Mine waste calcines were found to be stocked northeast of the abandoned open pit lake (Figure 2). The mine waste calcines have poor sorting, angular grains, and can reach around 6 to 8 m in thickness. The calcine exhibits a reddish to orangey brown color due to the weathered iron bearing minerals and residual cinnabar. It mostly contains crushed bedrock material hosting the Hg deposit, along with few grains of silica from the opalite, which brought on the mineralization.

In this study, the measured THg in mine waste calcines have concentrations ranging from 52.7–924.2 mg kg^{-1}. These calcines were stocked at the shore of the lake and scattered up to 500 m northeast of the lake. Total Hg concentrations found in the mine waste calcines are largely similar to the concentrations measured in the same study area in 2002–2003, but this time, a higher concentration (924.2 mg kg^{-1}) was measured at one sampling point. A possible cause of the high THg concentration in mine waste calcines, is the inefficient and incomplete calcination or roasting of cinnabar ores. The range of THg concentrations in mine waste calcines measured in this study were generally within the range of THg concentrations on calcines from China, New Zealand, and the USA, as presented in Table 1.

Total Hg concentrations measured in soil of Honda Bay wharf and stockpiled mine wastes exhibit similar ranges of concentration. Samples collected in Honda Bay wharf display the same physical characteristics as the calcines stockpiled near the pit lake. This confirms that the soil materials used in the construction of the wharf were the mine waste calcines from PQMI mine site. High THg concentration (731.6 mg kg^{-1}) was also measured in the calcines stocked near the wharf. It was higher than the estimated THg concentration (570 mg kg^{-1}) of all the calcine materials used in Honda Bay wharf in 1996 [24]. Sediment samples collected 10 m north of the wharf shore have Hg concentrations of >100 mg kg^{-1}, which is in agreement with the concentration reported by Benoit and co-workers [24] in 1996. As stated in previous studies [20,21,24], Honda Bay wharf is considered one of the possible sources of Hg contamination in Honda Bay. The removal of Honda Bay wharf as well as the dredging of contaminated tailings and sediments in Honda Bay, were suggested in the study of Williams and co-workers in 1996 [21], but these recommendations were not followed. Today, the inhabitants of the villages located in the vicinity of the abandoned mine area including Honda Bay, which is home to around 10,000 inhabitants, are exposed to potentially serious health issues due to Hg contamination.

Table 1. Total Hg concentrations in soils and mine waste calcines collected from selected Hg mines worldwide.

Location	n	Total Hg (mg kg^{-1})
Palawan Hg mine, Philippines (This study)		
Soils	31	0.04–67.5
Mine waste calcines	7	52.7–924.2
Palawan Hg mine, Philippines [11,22]		
Soils	21	0.012–168.72
Mine waste calcines	11	3.68–660
Wuchuan Hg mines, Guizhou, China [5]		
Soils	*	0.33–320
Mine waste calcines	*	79–710
East Guizhou Hg mines, China [6]		
Soils	*	5.1–790
Mine waste calcines	*	5.7–4400
West-central Nevada Hg mines, USA [8]		
Mine waste calcines	36	1.9–2000
Puhipuhi Hg mine, Northland, New Zealand [7]		
Soils	4	35.3–1486
Almaden Hg mine, Spain [15]		
Soils	6	13–64
Usagre Hg mine, Spain [4]		
Soils	26	0.2–311
La Soterraña Hg mine, Northern Spain [3]		
Soils	56	1.73–502
Branalamosa Hg mine, Northern Spain [3]		
Soils	28	0.5–895
Maramuñiz Hg mine, Northern Spain [3]		
Soils	23	2.0–577
Global background for pristine soils [25]		0.045–0.16

* number of samples was not provided in the literature.

3.2. THg in Sediment in Tagburos River and Pit Lake

Tagburos River, which is the main tributary in the abandoned mine area, has measured THg concentrations in the sediments ranging from 1.8–119 mg kg^{-1}. Sediments found in the river were influenced by the geology of the surrounding area, as the exposed soils with Hg can be transferred into the river and eventually deposited in Honda Bay. High Hg concentration in the river sediments suggest that exposed mine waste calcines with elevated Hg concentration in the abandoned mine area are being carried to the river by erosion and surface runoff, and are eventually deposited in Honda Bay where they become a source of Hg contamination and enter the food chain. Rehabilitation programs, such as revegetation, are recommended in this abandoned mine area, to avoid the erosion of soil particles with high Hg concentrations going to the river and depositing in Honda Bay.

High THg concentration measured in the pit lake sediments ranged from 37.5–454.1 mg kg^{-1}. The highest concentration was measured near the exposed bedrock on the lake shore, as shown in the inset of Figure 3. These values are within the range of the measured THg in pit lake sediments (6.9–400 mg kg^{-1}) in 2003 [11]. The high THg concentration at the lake bottom may be due to naturally occurring cinnabar in the bedrock, and deposition of eroded calcine stocked near the lakeshore.

3.3. THg in Soils

Natural concentrations of Hg in soil are dependent on the source parent rock's mercury concentration, which becomes concentrated through the processes of weathering and natural erosion. At the vicinity of abandoned mine area, measured THg concentrations (0.04–67.5 mg kg^{-1}) are 70 times higher than the global background for pristine soils (Table 2). The soil THg concentrations measured in this study were lower than the range of concentrations (0.01–168.7 mg kg^{-1}) analyzed in the same area in 2002–2003. Though the lowest Hg concentration measured was 0.04 mg kg^{-1}, corresponding to the background concentration; the other parts of the area might have been influenced by the movement of Hg-laden soil particles from the stocked mine waste calcines and other mine debris scattered in the abandoned mine area during operations. With consideration of the elevated sedimentation rate at Honda Bay (1 cm/year) [10], annual soil erosion to the bay via Tagburos River is likely a major source of Hg contamination for downstream ecosystems. Another human exposure pathway of Hg in the area is the dust coming from the bare soil during summer, which produces particulate-bound Hg that may be inhaled by the residents in the area.

Table 2. Mean total Hg concentration of different soil and sediment samples collected from different areas.

Site Location	n	Mean (Range) Concentration (mg kg^{-1})	S.D. (mg kg^{-1})	Global Background (mg kg^{-1}) [25]
Pit lake	11	124.5 (37.5–454.1)	123.1	0.045–0.16
Tagburos River	14	36.7 (1.8–119)	41.4	0.045–0.16
Honda Bay	28	3.5 (0.04–12.7)	3.6	0.045–0.16
Honda Bay wharf	6	357.3 (102–731.6)	245.5	-
Mine waste calcines	7	231.4 (52.7–924.2)	309.8	-
Other rivers	5	10.5 (5.3–11)	3.7	0.045–0.16
Soils	31	7.4 (0.04–67.5)	13.8	0.045–0.16

3.4. THg in Honda Bay Sediment

For the past four decades after the abandonment of the Hg mine operations by PQMI, there were very limited studies and monitoring conducted to determine the Hg concentrations in Honda Bay marine sediments. In the present study, analytical results of the THg concentrations in sediments collected from Honda Bay ranged from 0.04–12.7 mg kg^{-1}. This range of data shows a greater range of THg concentration, compared to the previous measurements conducted by different researchers in the same area some 20 years ago [20,21,24]. In the study by Benoit and co-workers in 1994 [21], the estimated Hg concentration in Honda Bay sediments (0.003–3 mg kg^{-1}) came from the wharf, which transported seaward or along the coast; while in this study, the identified possible sources of Hg contamination in Honda Bay marine sediments are the wharf (up to 12.7 mg kg^{-1}, 250 m from the wharf) and Tagburos River (up to 9.6 mg kg^{-1} in river mouth). The present THg concentration in the marine sediment also higher than the global Hg background concentration for sediment. Furthermore, the values gathered in the present study are in the same range of sediment THg concentration in the dredged area of Minamata Bay in 2006, as shown in Figure 4 [26]. Finally, such an elevated range of Hg concentrations at Honda Bay could also be biomethylated inside the Bay, becoming a significant source of methylmercury for both the marine and freshwater food chains that might be consumed by local inhabitants.

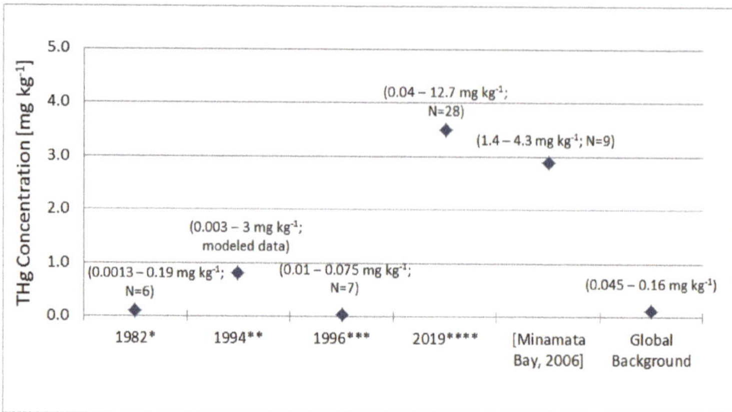

Figure 4. Mean and range of THg concentrations in surface marine sediments of Honda Bay, from 1982 to present. Data from Honda Bay were compared with data from the dredged Minamata Bay (2006) and global background. * Kapauan et al. [20], ** Benoit et al. [24], *** Williams et al. [21], **** This study.

3.5. *THg of Sediment from Other Rivers*

Collection of sediment samples were even extended to other rivers that flow through the abandoned mine area, to determine the trend of Hg concentration surrounding the study area. The measured THg concentrations across several rivers in the area ranged from 5.3–11 mg kg^{-1}. These values are more than 60 times higher than the global background, indicating Hg accumulation in the river caused by the erosion of the soil in the area with a naturally higher Hg background concentration. In comparison, sediments in Tagburos River show significantly higher Hg concentration than the other rivers since it traverses the abandoned mine site and the stockpiled calcines. Mercury may be moved along with the sediments and through water runoff, thereby increasing Hg values in the sediment deposited in Tagburos River.

4. Conclusions

In order to determine the pollution status and availability of Hg from the abandoned Hg mine in Puerto Princesa City, this study investigated the soil and sediment samples collected from the surroundings of the abandoned PQMI mine area. Results of the Hg analyses of these samples followed similar contamination patterns from other reported Hg mine sites worldwide. Mercury concentration in mine waste calcines ranged from 52.7–924.2 mg kg^{-1}, of which some values were higher than those analyzed in 2003. Mercury levels measured in Honda Bay wharf were found to be similar in Hg concentration and characteristic with the mine waste calcines, confirming that the construction of wharf utilized the calcines from the PQMI mine site. Sediment samples obtained from Tagburos River and other tributaries, Honda Bay, and soils also exhibited Hg values that exceeded the global background, due to naturally high Hg concentrations in the mineralized area, and from the erosion and deposition of mine waste calcines into the surrounding areas. A greater range of Hg concentrations were measured at Honda Bay compared to the previous measurements conducted by several researchers in the same area some 20 years ago. These increased Hg concentrations in Honda Bay sediments may undergo methylation and enter the food chain, which was observed in Hg mining sites around the world [3,7,12]. This study concludes that one of the pathways of the reported Hg contamination in the area is caused by sediment transport from two point sources: (1) the calcines stocked at the shore of the pit lake and nearby area with scattered calcine, and (2) Honda Bay wharf. Sediment transport from the first point source was facilitated by Tagburos River and its tributaries, while the second point source was due to coastal erosion. Further studies are needed to evaluate the methylmercury levels in biota and particle-bound Hg in the air to evaluate health issues. Mining rehabilitation activities, such as revegetation to reduce soil erosion and soil particle suspension, need to be implemented based on the results of this study.

Author Contributions: Conception and design of study and supervision, J.S. and C.R.G.; acquisition, analysis, interpretation of data, drafting the manuscript, and approval of the version of the manuscript to be published, J.S., C.R.G., A.T., and R.R. All authors have read and agreed to the published version of the manuscript.

Funding: This research was funded by the Philippines' Department of Science and Technology's Grants-in-Aid Program (DOST-GIA).

Acknowledgments: The authors acknowledge the Department of Science and Technology; the Philippine Council for Industry, Energy, and Emerging Technology Research and Development (DOST-PCIEERD); and the Provincial Science and Technology Center of Palawan.

Conflicts of Interest: The authors declare no conflict of interest.

References

1. Samaniego, J.O.; Gibaga, C.R.L.; Tanciongco, A.M.; Rastrullo, R.M.; Costa, M.A.V. Surface water characteristics in the vicinity of abandoned mercury mine site in Puerto Princesa City, Philippines. *Philipp. J. Sci.* **2019**, *148*, 493–498.
2. Selected Mercury Waste Hotspots around the World. Available online: https://ipen.org/sites/default/files/documents/mercury_waste_hotspots_world_map-en.pdf (accessed on 14 October 2019).
3. Ordoñez, A.; Alvarez, R.; Charlesworth, S.; De Miguel, E.; Loredo, J. Risk assessment of soil contaminated by mercury mining, Northern Spain. *J. Environ. Monit.* **2011**, *13*, 128–136. [CrossRef]
4. Higueras, P.; Lorenzo, S.; Esbrí, J.M.; García-Noguero, E.M.; Reyes-Bozo, L. Soil pollution related to mercury-mining activities in the proximity of Usagre (Badajoz, SW Spain). *Int. J. Min. Reclam. Environ.* **2014**, *28*, 377–388. [CrossRef]
5. Qiu, G.; Feng, X.; Wang, S.; Shang, L. Environmental contamination of mercury from Hg-mining areas in Wuchuan, northeastern Guizhou, China. *Environ. Pollut.* **2006**, *142*, 549–558. [CrossRef] [PubMed]
6. Qiu, G.; Feng, X.; Wang, S.; Shang, L. Mercury and methylmercury in riparian soil, sediments, mine-waste calcines, and moss from abandoned Hg mines in east Guizhou province, southwestern China. *Appl. Geochem.* **2005**, *20*, 627–638. [CrossRef]
7. Gionfriddo, C.M.; Ogorek, J.M.; Butcher, M.; Krabbenhoft, D.P.; Moreau, J.W. Mercury distribution and mobility at the abandoned Puhipuhi mercury mine, Northland, New Zealand. *N. Z. J. Geol. Geop.* **2015**, *58*, 78–87. [CrossRef]
8. Gray, J.E.; Crock, J.G.; Fey, D.L. Environmental geochemistry of abandoned mercury mines in West-Central Nevada, USA. *Appl. Geochem.* **2002**, *17*, 1069–1079. [CrossRef]
9. Irving, E.M. Note on the first discovery of commercial mercury in the Philippines. *Philipp. Geol.* **1954**, *9*, 1–8.
10. Williams, T.M.; Weeks, J.M.; Apostol, A.N.; Miranda, C.R. Assessment of mercury contamination and human exposure associated with coastal disposal of waste from a cinnabar mining operation, Palawan, Philippines. *Environ. Geol.* **1999**, *39*, 51–60. [CrossRef]
11. Gray, J.E.; Greaves, I.A.; Bustos, D.M.; Krabbenhoft, D.P. Mercury and methylmercury contents in mine-waste calcine, water, and sediment collected from the Palawan Quicksilver Mine, Philippines. *Environ. Geol.* **2003**, *43*, 298–307. [CrossRef]
12. Navarro, A.; Biester, H.; Mendoza, J.L.; Cardellach, E. Mercury speciation and mobilization in contaminated soils of the Valle del Azogue Hg mine (SE, Spain). *Environ. Geol.* **2006**, *49*, 1089–1101. [CrossRef]
13. Wiederhold, J.G.; Smith, R.S.; Siebner, H.; Jew, A.D.; Brown, G.E.; Bourdon, B.; Kretzschmar, R. Mercury isotope signatures as tracers for Hg cycling at the New Idria Hg mine. Environ. *Sci. Technol.* **2013**, *47*, 6137–6145. [CrossRef] [PubMed]
14. Yin, R.S.; Feng, X.B.; Meng, B. Stable mercury isotope variation in rice plants (Oryza sativa L.) from the Wanshan Mercury Mining District, SW China. Environ. *Sci. Technol.* **2013**, *47*, 2238–2245. [CrossRef]
15. Fernandez-Martınez, R.; Rucandio, I. Total mercury, organic mercury and mercury fractionation in soil profiles from the Almadén mercury mine area. *Environ. Sci.* **2014**, *16*, 333–340. [CrossRef]
16. Liu, J.; Shi, J.Z.; Yu, L.M.; Goyer, R.A.; Waalkes, M.P. Mercury in traditional medicines: Is cinnabar toxicologically similar to common mercurials? *Exp. Biol. Med.* **2008**, *233*, 810–817. [CrossRef]

17. Regnell, O.; Watras, C.J. Microbial mercury methylation in aquatic environments—A critical review of published field and laboratory studies. *Environ. Sci. Technol.* **2019**, *53*, 4–19. [CrossRef]
18. Harding, G.; Dalziel, J.; Vass, P. Bioaccumulation of methylmercury within the marine food web of the outer Bay of Fundy, Gulf of Maine. *PLoS ONE* **2018**, *13*. [CrossRef] [PubMed]
19. Bernhoft, R.A. Mercury toxicity and treatment: A review of the literature. *J. Environ. Public Health* **2012**, *2012*. [CrossRef]
20. Kapauan, A.F.; Kapauan, P.A.; Tan, E.O.; Verceluz, F. Total mercury in water and sediments from the Honda Bay area in Palawan. *Philipp. J. Sci.* **1982**, *111*, 135–144.
21. Williams, T.M.; Weeks, J.M.; Apostol, A.; Miranda, C. *Assessment of Toxicity Hazard Associated with Former Cinnabar Mining and Tailings Disposal in Honda Bay, Palawan, Philippines*; British Geological Survey, Overseas Geology Series Technical Report WC/96/31; Keyworth: Nottingham, UK, 1996.
22. Maramba, N.P.C.; Reyes, J.P.; Francisco-Rivera, A.T.; Panganiban, L.C.R.; Dioquino, C.; Dando, N.; Timbang, R.; Akagi, H.; Castillo, M.T.; Quitoriano, C.; et al. Environmental and human exposure assessment monitoring of communities near an abandoned mercury mine in the Philippines: A toxic legacy. *J. Environ. Manag.* **2006**, *81*, 135–145. [CrossRef]
23. Rappler Philippines. Available online: https://www.rappler.com/nation/172410-widespread-mercury-poisoning-puerto-princesa-villages-palawan (accessed on 10 September 2019).
24. Benoit, G.; Schwantes, J.M.; Jacinto, G.S.; Goud-Collins, M.R. Preliminary study of the redistribution and transformation of HgS from cinnabar mine tailings deposited in Honda Bay, Palawan, Philippines. *Mar. Pollut. Bull.* **1994**, *28*, 754–759. [CrossRef]
25. Siegel, F.R. *Environmental Geochemistry of Potentially Toxic Metals*; Springer: Berlin, Germany, 2002; pp. 18–23.
26. Tomiyasu, T.; Matsuyama, A.; Eguchi, T.; Fuchigami, Y.; Oki, K.; Horvat, M.; Rajar, R.; Akagi, H. Spatial variations of mercury in sediment of Minamata Bay, Japan. *Sci. Total Environ.* **2006**, *368*, 283–290. [CrossRef] [PubMed]

applied sciences

MDPI

Article

Particulate Mercury and Particulate Organic Matter in the Itenez Basin (Bolivia)

Fabiola Guzmán-Uria [1], Isabel Morales-Belpaire [2,*], Dario Achá [1,*] and Marc Pouilly [3]

1 Unidad de Calidad Ambiental (UCA), Instituto de Ecología, Campus Universitario de Cota Cota, Universidad Mayor de San Andrés, La Paz, Bolivia; fabiolainesguzman@gmail.com

2 Instituto de Biología Molecular y Biotecnología, Campus Universitario de Cota Cota, Universidad Mayor de San Andrés, La Paz, Bolivia

3 IRD–Laboratoire BOREA (MNHN, CNRS, IRD, Universités Paris Sorbonne, Caen et Antilles), 61 rue Buffon, 75005 Paris, France; marc.pouilly@ird.fr

* Correspondence: imorales@fcpn.edu.bo (I.M.-B.); dacha@fcpn.edu.bo (D.A.)

Received: 31 October 2020; Accepted: 23 November 2020; Published: 26 November 2020

Abstract: In rivers and other freshwater bodies, the presence of mercury can be due to direct contamination by anthropic activities such as gold mining. However, it can also be attributed to atmospheric deposition and erosion, runoff, or lixiviation from surrounding soils. In the case of the Amazon rainforest, high mercury contents have been reported for litter and topsoil, which could affect the mercury concentrations in water bodies. Samples of suspended particulate matter were obtained from a transect of the Itenez River, associated lakes, and some of its tributaries. The aim was to obtain information on particulate mercury's origin in the study area and determine the relationship between particulate mercury and particulate organic carbon. The concentration of mercury, organic matter, and the C:N ratio of the suspended matter was determined. The concentration of particulate mercury by water volume depended on changes in suspended matter loads, which in turn were mostly affected by the nature of the watershed or sediment resuspension. The observed values for the percentage of organic matter and the C:N ratio suggest that most of the mercury content in rivers and lakes originated from soils. A positive correlation was found between mercury concentration by weight of particulate matter and organic carbon content in particles. This correlation might be due to the direct binding of mercury to organic matter through functional groups like thiols or to an indirect effect of oxyhydroxides that can adsorb mercury and are associated with organic matter.

Keywords: particulate mercury; suspended particulate matter; particulate organic carbon; Amazon rainforest

1. Introduction

Mercury (Hg) is a significant environmental and health issue in the Amazon basin. Mercury's toxicity is particularly elevated due to the property of its organic form, methylmercury, to penetrate biological organisms and to generate accumulation during the individual's life and magnification along the trophic chain. In tropical forests, typical vegetation with broad leaves facilitates the capture of mercury from atmospheric deposition. Consequently, high contents of this metal have been reported for forest litter and topsoil [1]. In water bodies, mercury content can be related to atmospheric deposition, but most originates from the runoff or lixiviation of superficial and subterraneous water from mercury-laden soils [2].

The influx of mercury to water bodies from runoff can be exacerbated by natural erosion and anthropic activities such as deforestation, agriculture, mining, or urbanization. In rivers that drain deforested watersheds, an increase in total mercury in the water column and surface sediments has been observed and attributed to the erosion of fine particles from soils [3,4]. Additionally, forest fire for land conversion can give rise to an increase in Hg mobility [5]. However, the impact of deforestation on mercury concentrations in rivers and lakes varies in different watersheds [6].

Gold mining activities are also responsible for the discharge of mercury to the environment [6]. Exogenous (industrial) mercury used for gold mining can be emitted to the atmosphere or released into water bodies in the form of droplets and mercury-coated sediments [7]. Pollution may be worst when cyanide is used as part of the process because cyanomercuric complexes are more easily transported [7]. Schudel et al. [8], using stable isotopes of mercury, showed that mercury from mines where cyanide was used could be traced 120 km downstream. Although gold mining causes deforestation in smaller areas than cattle rising or agriculture, it is one of the leading causes of deforestation in South America [9]. Moreover, informal gold mining can cause more substantial soil perturbation than other activities due to the removal of the organic layer under the trees, which increases the loads of sediments containing high mercury concentrations [10].

Pouilly and coworkers [11] reported data on mercury concentrations in fish populations from different trophic levels in the Itenez basin (Bolivia). They also reported average values for particulate mercury in Blanco, San Martín, and Itenez rivers and their associated floodplain lakes. The authors highlighted that no clear relationship between particulate mercury concentrations in water and mercury concentrations in fish muscle was observed, indicating that bioavailability must be taken into account for interpreting the results. A second study by Pouilly et al. [12] suggested that periphyton's contribution to the food web could explain higher mercury bioavailability since it is a site of mercury methylation [13]. We report additional data to those that were presented by Pouilly et al. [11] about suspended particulate matter, total particulate mercury, particulate organic carbon, and the C/N ratio. We took into account some additional sampling points directly influenced by deforestation and mining and performed a seasonal analysis (early and late dry season). These additional data were analyzed to obtain information on the origin of particulate mercury in the study area and determine the relationship between particulate mercury and particulate organic carbon.

The data that we report and analyze correspond to samples that were taken from the Itenez basin in June and September 2007. No previous or more recent data on particulate mercury in the Itenez basin has been reported to our knowledge, although this river is one of the main tributaries of the upper Madeira River and its basin covers 303,300 km^2 shared between Brazilian and Bolivian territories [11]. The punctual values on particulate mercury that we report are not expected to reflect the current situation. However, the information on its relationship with anthropic activities and with organic matter contributes to the characterization of mercury pollution in that particular region. Moreover, this characterization is valuable for newer studies on mercury pollution in the region, particularly given recent deforestation, which has been aggravated by the 2019 and 2020 forest burnings.

2. Materials and Methods

Water samples were taken in June (early dry season) and September (late dry season) 2007 from a longitudinal transect in the Itenez River, which is located at the frontier between Bolivia and Brasil and from the tributaries Blanco and San Martín. The sampled transect in Itenez river is 660 km long, starting at the "Piso Firme" locality (S 13°31′36.9″ W 61°49′40.7″) and finishing 100 km upstream of the confluence with Mamoré River (S 12°11′43.0″ W 64°40′00.5″). Floodplain lakes related to the Itenez, San Martin, and Blanco Rivers were also sampled. Additional samples were obtained from three streams that cross Bolivian territory influenced by the "San Simon" goldmine and from three tributaries of Itenez that cross highly deforested Brazilian territory. The location of the sampling points can be observed in Figure 1. A more detailed description of the study area can be found in [11].

Five-liter plastic containers were used for sampling. These were conditioned by following a slight modification of the protocol proposed by Montgomery et al. [14]. The containers were first washed with tap water and metal-free detergent and rinsed with MilliQ water. After this, they soaked overnight with a 7% NaOH solution. They were then rinsed with MilliQ water and soaked with 10% HCl for an additional 12 h. Finally, they were rinsed three times with MilliQ water. Once clean, the containers were kept in hermetic plastic bags until their use. For the sampling of SPM, GF/F (0.7 μm) filters were used. Before sampling, the filters were placed in packages wrapped with aluminum foil and heated up to 500 °C for 8 h. Upon cooling, the filters were weighted and individually placed in Petri dishes.

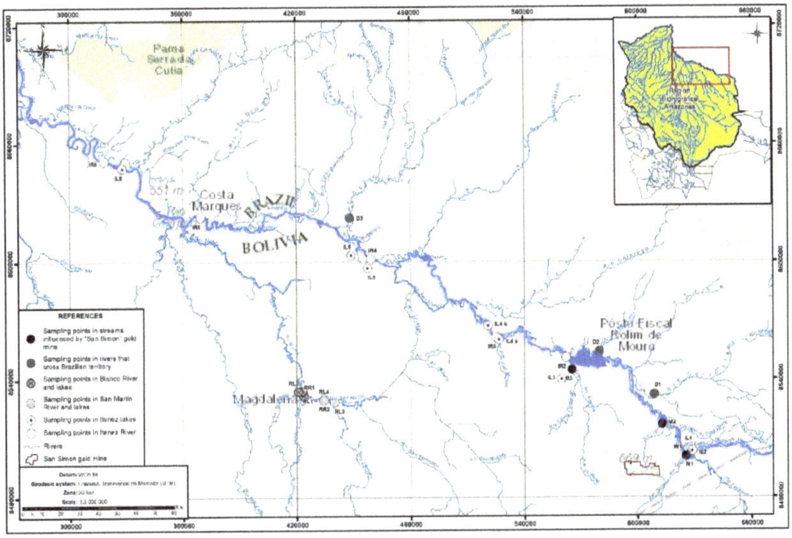

Figure 1. Location of sampling points.

Water was collected from the mainstream of rivers and the center of lakes by submerging the containers around 1 m deep. Following [13], the person in charge of sampling wore clean gloves throughout the procedure. The containers were rinsed three times with the river or lake water before being filled. Additionally, in each sampling point, temperature (°C), electric conductivity, and pH were measured with CONSORT C353 multimeter (Belgium). The depth of water was also registered.

SPM was obtained in the field by vacuum filtration of the water sample. The volume of water that caused saturation of the filter was annotated. Each filter was then placed in a Petri dish and stored at −17 °C until analysis.

Mercury content of SPM was determined at the Laboratorio de Calidad Ambiental of the University of San Andrés (La Paz, Bolivia). The filters that contained SPM were oven-dried at 40 °C for 24 h and weighted. Each filter was then introduced in a thermoresistant screw-capped 50 mL glass tube. 1 mL of nitric acid and 100 μL of hydrochloride acid were added to the tubes, and these were sealed with Teflon and capped. All tubes were heated at 100 °C for two hours. After cooling, 200 μL of hydrogen peroxide was added, and the tubes were again heated at 100 °C for two hours. 5 mL of ultrapure water was finally added.

Mercury of the digested samples was quantified with Cold Vapor Atomic Fluorescence Spectrometer (CVAFS, Brooks Rand Model III, Seattle, WA, USA) using the EPA 1631 Revision E: (EPA 2002) protocol. Reference samples were obtained from the National Research Council of Canadá: TORT-2 lobster hepatopancreas and MESS-3 Marine Sediment Reference Materials for Trace Metals (NRC-CNRC 2016). Total mercury concentration in SPM was expressed by water volume ($[Hg]_v$) and SPM weight ($[Hg]_W$). SPM from filters was additionally analyzed for particulate organic carbon (POC) and nitrogen in the Laboratorio de Medios Analíticos Cayenne- IRD in French Guyana with the CHNS THERMOQUEST AS2100 analyzer.

Wilcoxon test was applied to compare data obtained among sites (rivers and lakes) and seasons (early and late dry season). Spearman correlation coefficients between variables were calculated. All analyzes were carried out with R statistical computing freeware program (http://www.r-project.org).

3. Results and Discussion

A significant decrease in water depth was observed in September (late dry season, Tables S1 and S2) for most sampling points, except IR 1 and D3 (Tables S1 and S2). Most physicochemical characteristics did not vary among sites or seasons (Tables S1 and S2). The Blanco River showed a higher conductivity than the other water bodies (50 μS cm^{-1} as compared to a mean of 27 ± 4 μS cm^{-1} for the Itenez River and 20 μS cm^{-1} for San Martin), and this value increased during the late dry season (80 μS cm^{-1}).

Quantities of SPM showed a mean value of 5.6 ± 1.1 mg/L with minor variations along the Itenez river in June (Table S1). In September, the quantities of SPM along the Itenez River were significantly higher ($p = 0.005$) than in June (Figure 2A). For point IR6 (Vuelta Grande), the value obtained in September more than doubled the value measured in June (12.4 mg/L in September compared to 5.7 mg/L in June, Table S1). Such an increase can be related to the resuspension of sediments by physical factors or bioturbation since the water column at this point became very shallow (0.9 m) [15].

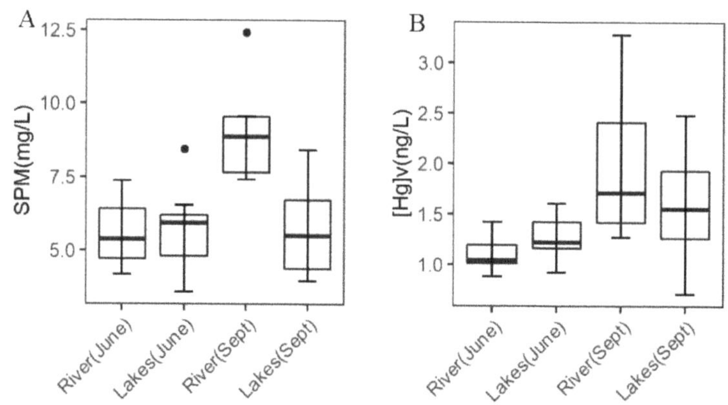

Figure 2. Comparison of suspended particulate matter (**A**) and particulate mercury concentration by volume (**B**) for the Itenez River and associated lakes in the two seasons.

In June, SPM values for the Itenez lakes were similar to those obtained for the river (Figure 2A); this is in agreement with the fact that, during the high water level season, rivers and lakes are connected. No significant difference for SPM in the Itenez lakes was observed between September and June, except for lake IL9, which showed a strong increase of suspended matter (from 4.8 mg/L in June to 100.2 mg/L in September, Table S1). In September, the connections between the river and the lakes were interrupted, and most floodplain lakes of Itenez showed slightly lower concentrations of SPM than the river, probably because of a sedimentation process and the fact that they did not receive new particulate material. Lake IL9 became very shallow in September (1.70 m), and resuspension phenomena can explain the very high value of SPM (Table S1).

Two of the sampled tributaries of Itenez, a river from Brazilian territory that crosses highly deforested areas and the Colorado stream, which is influenced by gold mining showed in June the highest values for SPM (20.7 mg/L and 15.2 mg/L, respectively) (Table S2). However, the other tributaries directly influenced by anthropic activities showed values similar to those of the Itenez river. In September, the Asunta Stream showed a very high value (36.7 mg/L) for SPM, but the other two streams influenced by the "San Simon" goldmine had low values (6.4 and 9.7 mg/L). On the other hand, in June, Blanco river, which drains white waters, contained higher amounts of particulate matter than Itenez river (27.7 mg/L), while San Martín, a clear water river, and its lakes, showed the lowest values (2.9 mg/L, Table S2). Both rivers cross territory with little anthropic influence [11]. In September, however, concentrations of SPM in San Martin increased strongly (up to 19.9 mg/L), probably due to the resuspension of sediments [15], while this was not the case for Blanco. Although deforestation and mining give rise to disturbance of the upper horizons of soil and cause an increase in the loads of solid particles that reach water bodies [9,10], factors not related to anthropic intervention must also be taken into account in order to determine the causes of high SPM.

The concentration of particulate mercury by water volume ($[Hg]_V$) followed the same pattern as particulate suspended matter, showing little variations along the Itenez river (mean value 1.10 ± 0.19 ng/L) and its associated lakes in June (mean value 1.26 ± 0.23 ng/L, see Table S1). In September, the Itenez river had significantly higher $[Hg]_V$ than in June (p = 0.0129) (Figure 2B), as was observed for SPM.

Although high values for $[Hg]_V$ were measured in June for one of the rivers coming from deforested Brazilian territory and for the Colorado stream (3.83 ng/mL and 2.45 ng/L), the other tributaries influenced by anthropic activities showed concentrations of particulate mercury similar to those of Itenez river (Table S2). The lowest values for $[Hg]_V$ were observed in June for the San Martín River and its floodplain lakes (around 0.70 ng/L), while the highest value of all samples was obtained in the Blanco River (3.96 ng/L, Table S2). In September, the Asunta stream showed a substantial increase in $[Hg]_V$ (7.00 ng/L), but this was not observed for the other tributaries influenced by anthropic activities (Table S2).

The samples were taken from a very large area with different degrees of anthropic intervention, which goes from no intervention to gold-mining and deforestation. This area includes rivers and floodplain lakes, most of which contain clear water while some contain white water. No data are available for comparisons in this region, and there are few data on particulate mercury under similar conditions in the Amazonian basin. One exception is data from Roulet et al. [16] for suspended particulate matter and particulate mercury in the Tapajos river. The Tapajos river and the Itenez river are considered clear water rivers [16]. The values reported by Roulet et al. [17] for fine particulate matter and particulate mercury in the early dry season are similar to our data for the Itenez river and its lakes. The increase of SPM that we observed in the dry season (September) was also observed at Tapajos and was attributed to the decrease of water volumes. However, in our work, mercury associated with the coarse and fine particulate matter has been pooled, while Roulet et al. [17] did not report mercury concentration associated with coarse particulate matter. Comparisons must, therefore, be interpreted carefully.

More than the punctual values on particulate mercury by water volume, we were interested in the relationships between mercury and the degree and type of anthropic interventions. The relative contribution of mining activities and deforestation to mercury pollution in water bodies is still controversial and might depend on regional characteristics. Obritz et al. [18] indicate that the accumulation of mercury in soils depends on soil properties. Moreover, the accumulation in upper organic layers is different from the accumulation in mineral horizons. This implies that the contribution of soil mercury to water bodies can vary between regions. They also point out that runoff from contaminated sites such as gold mines can dominate Hg loads in water bodies. Horbe et al. [19] indicate that there are large variations in geochemical values of mercury in western Amazonia, and, by then, the contribution of different soils to Hg loads in rivers and lakes is variable. In our case, no systematic increase in $[Hg]_V$ was observed in sample points of Itenez river nearest to the mining sites or in streams affected by the "San Simon" mine. Furthermore, the obtained values for Hg loads in rivers that crossed highly deforested Brazilian territory and for rivers that crossed territories with little anthropic intervention showed no clear relationship between anthropic activities and particulate mercury loads. Phenomena such as resuspension, soil properties, and the water properties that the river drains (clear/white) must be taken into account when interpreting Hg loads.

When the concentration of total particulate mercury was evaluated by weight of SPM ($[Hg]_W$), the values obtained in June for the complete set of samples fluctuated between 133 and 327 ng g^{-1} (Figure 3A, Tables S3 and S4). No significant variation between June and September was observed for the Itenez river or its lakes. In samples from tributaries affected by mining activities and rivers that cross highly deforested Brazilian territory, the values for $[Hg]_W$ were similar to those of Itenez river. Such was also the case for samples from Blanco and San Martin rivers.

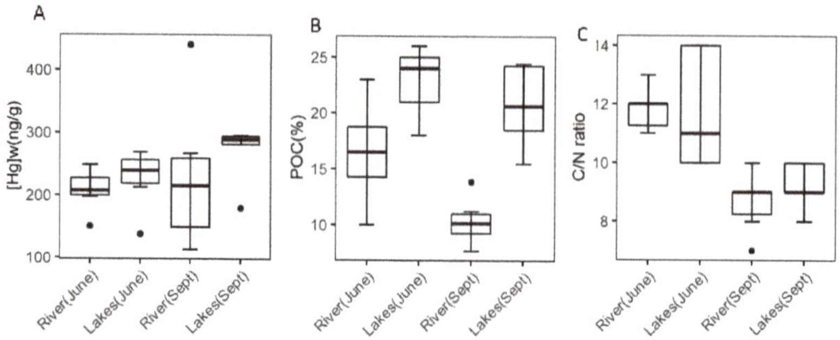

Figure 3. Characteristics of suspended particulate matter including the concentration of mercury by weight ($[Hg]_W$) (**A**), organic carbon (**B**), and C/N ratio (**C**) for the Itenez River and associated lakes in June (early dry season) and September (late dry season).

The percentage of particulate organic carbon (POC) in SPM in all samples from the Itenez basin varied from 6 to 26% in June (Tables S3 and S4). Samples from Itenez lakes tended to contain higher POC percentages than those from the Itenez river. POC decreased significantly ($p = 0.02$) in September in Itenez River, but no significant variation was observed for its lakes. Lower percentages of organic carbon associated with suspended matter can be due to lower organic matter inputs to water bodies in dry months and carbon turnover. Although particulate organic carbon is less bioavailable than dissolved carbon, some carbon turnover takes place [20].

In June, the C:N ratio for particulate matter from the Itenez river, lakes, and tributaries showed values between 9.9 and 14.9. In September, the C:N ratio values were lower, fluctuating between 7.4 and 11.2 for all samples (Tables S3 and S4). This decrease was significant for both the Itenez river ($p = 0.004$) and its lakes ($p = 0.009$), Figure 3C, and it can be related to preferential sorption of nitrogen compounds, as observed by Aufdenkampe et al. [21]. For both June and September, and for both habitats (lakes and rivers), the obtained C:N values are characteristic of suspended matter originated from soil [17,22]. This finding implies that the particulate organic carbon and the particulate mercury are mostly terrigenous in our study area. Terrigenous POC has a lower tendency to drive mercury methylation than autochthonous material [23], suggesting relatively low bioavailability of mercury.

A significant negative correlation was observed between SPM and POC for June ($\rho = -0.739$, $p < 0.001$) and September (($\rho = -0.739$, $p < 0.001$). Likewise, SPM and $[Hg]_W$ were negatively correlated in June ($\rho = -0.640$, $p < 0.001$) and September ($\rho = -0.709$, $p < 0.001$, see also Figure 4A,B). Maurice-Bourgoin et al. [24] have also observed a negative correlation between SPM and $[Hg]_W$ for samples from the Negro, Solimoes, Amazon, and Madeira rivers. This "dilution effect" can be due to several processes. Higher loads of SPM generally imply that water also contains higher concentrations of colloids, which are not retained by filters but could absorb part of the available mercury [25]. Hsu-Kim et al. [6] indicate that there can be a "growth dilution" of particulate mercury caused by an increase in biomass attached to particles. On the other hand, for sampling points where strong resuspension of sediments was observed, POC's values and $[Hg]_W$ were much lower than in the other points. Sediments generally contain less organic carbon and mercury than suspended matter in the water column [3,23]. This difference can be related to high heterotrophic activity in sediments at the benthic and hyporheic zone [26]. High heterotrophic activity causes carbon turnover and can also cause mercury transformations, which leads to its separation from particles. A combination of the "dilution effect" and sediment resuspension could explain the behavior observed in Figure 4A,B.

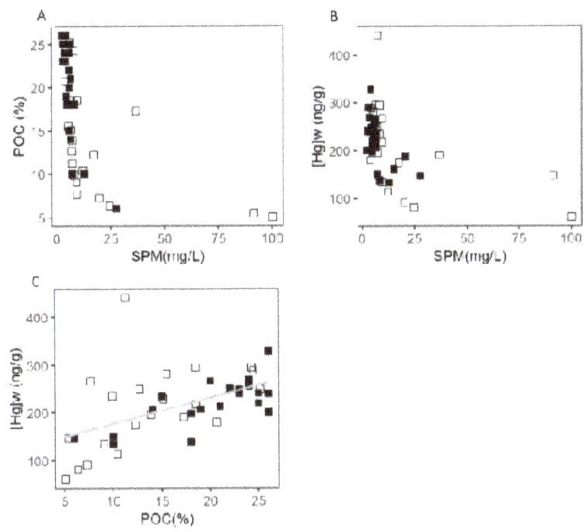

Figure 4. Relationships between particulate organic carbon (POC) and suspended particulate matter (SPM) (**A**), $[Hg]_W$ and SPM (**B**) and $[Hg]_W$ and POC (**C**) for samples taken in June (closed squares) and September (open squares).

Particulate organic carbon was positively correlated with particulate mercury (see Figure 4C) both in June ($\rho = 0.595$, $p = 0.004$) and September ($\rho = 0.595$, $p = 0.004$). Such a positive relationship was also reported by other authors [23,24], although in some other cases, no significant correlation was found [15,17]. The nature of the samples of the particulate matter probably influences this correlation. Roulet et al. [17] found no correlation between $[Hg]_W$ and the POC percentage in fine particulate matter. The difference between this result and ours might be due to the fact that, in our case, we included coarse material in the analysis; this coarse material could contain higher percentages of organic matter. The characteristics of the water that is drained by the river could also play a role in the relationship between $[Hg]_W$ and POC. Maia et al. [15] found no significant correlation between $[Hg]_W$ and POC for floodplain lakes of the Amazonian river. However, these lakes contained water classified as white or black water, while the Itenez river is considered a clear water river. The fact that not all authors find a positive correlation between particulate mercury and particulate organic matter can also be due to regional differences and types of soil. Therefore, our data can contribute to a wider analysis of which conditions give rise to such a positive relationship.

Samples were taken in two seasons of the same year, at the beginning of the low water season and by the end of the low water season. The relationships between suspended particulate matter, particulate organic matter, and particulate mercury by weight $[Hg]_W$ did not show seasonal variations. The possible terrigenous nature of particles could explain the seasonal invariance of these relationships

The nature of the bonds between mercury and particulate suspended matter could affect mercury's fate and the possibilities of its methylation. Our results suggest that particulate mercury binds to organic carbon. However, mercury adsorption has also been related to aluminum and iron oxyhydroxides [3,27], which, in turn, are closely associated with organic matter [17]. Therefore, further studies are necessary to establish if the correlation of particulate mercury with POC is due to direct binding of mercury with organic matter or binding with the oxyhydroxides associated with this organic matter. Moreover, several studies have shown that mercury binding to organic matter is mostly through thiols, sulfides, and disulfides [25,28]. Consequently, the nature of the particulate organic matter might influence the amount of captured mercury. A study on sediment cores taken from the Titicaca lake has shown that capturing mercury on solids depends more on quality than quantity of organic matter. There is a positive correlation between bound mercury and reduced sulfur content [29]. Quantification of reduced sulfur in SPM could help establish the nature of the bonds between mercury and organic matter.

4. Conclusions

The concentration of particulate mercury by water volume $[Hg]_v$ in the studied water bodies depended on phenomena that caused changes in suspended matter loads. The effect of anthropic activities on these loads could not be established since the nature of the watershed or resuspension had a more direct effect on concentrations of SPM.

The POC values and the C:N ratio observed in suspended particulate matter suggest that most of this material in Itenez, Blanco, and San Martin rivers and lakes is terrigenous for the two sampling dates considered (early and late dry seasons), which implies that particulate mercury in these water bodies mostly comes from soils. At the same time, particulate organic matter (SPM) and $[Hg]_W$ were negatively correlated with the quantity of suspended matter, which may suggest a dilution effect.

A positive correlation was found between $[Hg]_W$ and POC. Although this suggests that mercury is bound to organic matter, further studies are needed to clarify this since organic matter could be associated with oxyhydroxides, and these minerals contain binding sites for mercury.

Supplementary Materials: The following are available online at http://www.mdpi.com/2076-3417/10/23/8407/s1, Table S1. Characteristics of samples from Itenez river and its floodplain lakes, Table S2. Characteristics of samples from Blanco, San Martín, and tributaries of Itenez, Table S3. Contents of mercury ($[Hg]_W$), organic matter (POC), and C/N ratio in suspended particulate matter of Itenez river and floodplain lakes, Table S4. Contents of mercury ($[Hg]_W$), organic matter (POC) and C/N ratio in suspended particulate matter of Blanco, San Martin, and other tributaries of Itenez.

Author Contributions: Investigation, F.G.-U.; writing—review and editing, I.M.-B., review and editing, D.A., methodology, review, and editing, M.P. All authors have read and agreed to the published version of the manuscript.

Funding: This work was funded by the IRD (http://www.ird.fr) project Trophic Ecology of Amazonian Aquatic Systems (JEAI-EMAA). It also received financial support from a WWF Bolivia (http://bolivia.panda.org) funded project (KN10 WWF-IRD agreement). Published with the support of IRD through the (JEAI-TITICACA) project.

Acknowledgments: We are particularly grateful for all the advice and assistance given by Jean Louis Duprey for all the laboratory work. We thank the Trinidad WWF team, Itenez departmental park (PD-AMNI Ite nez), and Beni Prefecture (Bolivia) for the logistical support and all the people who helped in the field work, especially the fishermen and boat pilots: H. Ribero Rodriguez, T. Suarez, and J. Vasquez. We greatly appreciated the warm welcome in the Itenez communities of Bella Vista, Remanso, Mategua, and Versalles.

Conflicts of Interest: The authors declare no conflict of interest. The funders had no role in the design of the study; in the collection, analyses, or interpretation of data; in the writing of the manuscript, or in the decision to publish the results.

References

1. Teixeira, D.C.; Lacerda, L.D.; Silva-Filho, E.V. Mercury sequestration by rainforests: The influence of microclimate and different successional stages. *Chemosphere* **2017**, *168*, 1186–1193. [CrossRef] [PubMed]
2. Fitzgerald, W.F.; Lamborg, C.H. 9.04-Geochemistry of Mercury in the Environment. In *Treatise on Geochemistry*; Holland, H.D., Turekian, K.K., Eds.; Pergamon: Oxford, UK, 2007; pp. 1–47. ISBN 978-0-08-043751-4.
3. Roulet, M.; Lucotte, M.; Canuel, R.; Farella, N.; Courcelles, M. Increase in mercury contamination recorded in lacustrine sediments following deforestation in the central Amazon. *Chem. Geol.* **2000**, *165*, 243–266.
4. Molina, C.I.; Pouilly, M. Fuentes primarias de mercurio en Bolivia. In *Mercurio en Bolivia: Línea Base de Usos, Emisiones y Contaminación 2014*; Ministerio de Relaciones Exteriores & Ministerio de Medio Ambiente y Agua: La Paz, Bolivia, 2014; pp. 43–75.
5. Farella, N.; Lucotte, M.; Davidson, R.; Daigle, S. Mercury release from deforested soils triggered by base cation enrichment. *Sci. Total Environ.* **2006**, *368*, 19–29. [CrossRef] [PubMed]
6. Hsu-Kim, H.; Eckley, C.S.; Achá, D.; Feng, X.; Gilmour, C.C.; Jonsson, S.; Mitchell, C.P.J. Challenges and opportunities for managing aquatic mercury pollution in altered landscapes. *Ambio* **2018**, *47*, 141–169. [CrossRef] [PubMed]
7. Telmer, K.H.; Veiga, M.M. World emissions of mercury from artisanal and small scale gold mining. In *Mercury Fate and Transport in the Global Atmosphere*; Mason, R., Pirrone, N., Eds.; Springer US: Boston, MA, USA, 2009; pp. 131–172. ISBN 978-0-387-93957-5.

8. Schudel, G.; Miserendino, R.A.; Veiga, M.M.; Velasquez-López, P.C.; Lees, P.S.J.; Winland-Gaetz, S.; Davée Guimarães, J.R.; Bergquist, B.A. An investigation of mercury sources in the Puyango-Tumbes River: Using stable Hg isotopes to characterize transboundary Hg pollution. *Chemosphere* **2018**, *202*, 777–787. [CrossRef] [PubMed]
9. Alvarez-Berríos, N.L.; Mitchell Aide, T. Global demand for gold is another threat for tropical forests. *Environ. Res. Lett.* **2015**, *10*, 014006. [CrossRef]
10. Asner, G.P.; Tupayachi, R. Accelerated losses of protected forests from gold mining in the Peruvian Amazon. *Environ. Res. Lett.* **2016**, *12*, 094004. [CrossRef]
11. Pouilly, M.; Pérez, T.; Rejas, D.; Guzman, F.; Crespo, G.; Duprey, J.-L.; Guimarães, J.-R.D. Mercury bioaccumulation patterns in fish from the Iténez river basin, Bolivian Amazon. *Ecotoxicol. Environ. Saf.* **2012**, *83*, 8–15. [CrossRef]
12. Pouilly, M.; Rejas, D.; Pérez, T.; Duprey, J.-L.; Molina, C.I.; Hubas, C.; Guimarães, J.-R.D. Trophic Structure and Mercury Biomagnification in Tropical Fish Assemblages, Iténez River, Bolivia. *PLoS ONE* **2013**, *8*, e65054. [CrossRef]
13. Guimarães, J.R.; Meili, M.; Hylander, L.D.; de Castro e Silva, E.; Roulet, M.; Mauro, J.B.; de Lemos, R. Mercury net methylation in five tropical flood plain regions of Brazil: High in the root zone of floating macrophyte mats but low in surface sediments and flooded soils. *Sci. Total Environ.* **2000**, *261*, 99–107. [CrossRef]
14. Montgomery, S.; Lucotte, M.; Pichet, P.; Mucci, A. Total dissolved mercury in the water column of several natural and artificial aquatic systems of Northern Quebec (Canada). *Can. J. Fish. Aquat. Sci.* **1995**, *52*, 2483–2492. [CrossRef]
15. Maia, P.D.; Maurice, L.; Tessier, E.; Amouroux, D.; Cossa, D.; Pérez, M.; Moreira-Turcq, P.; Rhéault, I. Mercury distribution and exchanges between the Amazon River and connected floodplain lakes. *Sci. Total Environ.* **2009**, *407*, 6073–6084. [CrossRef] [PubMed]
16. Sioli, H. The Amazon and its main affluents: Hydrography, morphology of the river courses, and river types. In *The Amazon: Limnology and Landscape Ecology of a Mighty Tropical River and Its Basin*; Sioli, H., Ed.; Monographiae Biologicae; Springer: Berlin/Heidelberg, Germany, 1984; pp. 127–165. ISBN 978-94-009-6542-3.
17. Roulet, M.; Lucotte, M.; Canuel, R.; Farella, N.; De Freitos Goch, Y.G.; Pacheco Peleja, J.R.; Guimarães, J.-R.D.; Mergler, D.; Amorim, M. Spatio-temporal geochemistry of mercury in waters of the Tapajós and Amazon rivers, Brazil. *Limnol. Oceanogr.* **2001**, *46*, 1141–1157. [CrossRef]
18. Obrist, D.; Kirk, J.L.; Zhang, L.; Sunderland, E.M.; Jiskra, M.; Selin, N.E. A review of global environmental mercury processes in response to human and natural perturbations: Changes of emissions, climate, and land use. *Ambio* **2018**, *47*, 116–140. [CrossRef] [PubMed]
19. Horbe, A.M.C.; da Costa Lima, C.B.; Garnier, J. Factors driving mercury variability and background values in a tropical region: The case of western Amazonia. *J. S. Am. Earth Sci.* **2019**, *95*, 102279. [CrossRef]
20. Worrall, F.; Burt, T.P.; Howden, N.J.K.; Hancock, G.R.; Wainwright, J. The fate of suspended sediment and particulate organic carbon in transit through the channels of a river catchment. *Hydrol. Process.* **2018**, *32*, 146–159. [CrossRef]
21. Aufdenkampe, A.K.; Hedges, J.I.; Richey, J.E.; Krusche, A.V.; Llerena, C.A. Sorptive fractionation of dissolved organic nitrogen and amino acids onto fine sediments within the Amazon Basin. *Limnol. Oceanogr.* **2001**, *46*, 1921–1935. [CrossRef]
22. Amelung, W.; Zech, W.; Zhang, X.; Follett, R.F.; Tiessen, H.; Knox, E.; Flach, K.-W. Carbon, Nitrogen, and Sulfur Pools in Particle-Size Fractions as Influenced by Climate. *Soil Sci. Soc. Am. J.* **1998**, *62*, 172–181. [CrossRef]
23. Maia, P.D.; Maurice, L.; Tessier, E.; Amouroux, D.; Cossa, D.; Moreira-Turcq, P.; Etcheber, H. Role of the floodplain lakes in the methylmercury distribution and exchanges with the Amazon River, Brazil. *J. Environ. Sci.* **2018**, *68*, 24–40. [CrossRef]
24. Maurice-Bourgoin, L.; Quemerais, B.; Moreira-Turcq, P.; Seyler, P. Transport, distribution and speciation of mercury in the Amazon River at the confluence of black and white waters of the Negro and Solimões Rivers: Mercury in the amazon river at the confluence of black and white waters. *Hydrol. Process.* **2003**, *17*, 1405–1417. [CrossRef]
25. Liu, G.; Li, Y.; Cai, Y. Adsorption of Mercury on Solids in the Aquatic Environment. In *Environmental Chemistry and Toxicology of Mercury*; Liu, G., Cai, Y., O'Driscoll, N., Eds.; John Wiley & Sons, Inc.: Hoboken, NJ, USA, 2011; pp. 367–387. ISBN 978-1-118-14664-4.

26. Fischer, H.; Wanner, S.C.; Pusch, M. Bacterial abundance and production in river sediments as related to the biochemical composition of particulate organic matter (POM). *Biogeochemistry* **2020**, *61*, 37–55. [CrossRef]
27. Roulet, M.; Lucotte, M.; Canuel, R.; Rheault, I.; Tran, S. Distribution and partition of total mercury in waters of the Tapaj'os River Basin, Brazilian Amazon. *Sci. Total Environ.* **1998**, *213*, 203–211. [CrossRef]
28. Skyllberg, U. Chapter 13—Mercury Biogeochemistry in Soils and Sediments. In *Developments in Soil Science*; Singh, B., Gräfe, M., Eds.; Synchrotron-Based Techniques in Soils and Sediments; Elsevier: Amsterdam, The Netherlands, 2010; Volume 34, pp. 379–410.
29. Guédron, S.; Audry, S.; Acha, D.; Bouchet, S.; Point, D.; Condom, T.; Heredia, C.; Campillo, S.; Baya, P.A.; Groleau, A.; et al. Diagenetic production, accumulation and sediment-water exchanges of methylmercury in contrasted sediment facies of Lake Titicaca (Bolivia). *Sci. Total Environ.* **2020**, *723*, 138088. [CrossRef] [PubMed]

applied sciences

MDPI

Article

Impact of Old and Recent Gold Mining Sites on Mercury Fluxes in Suspended Particulate Matter, Water and Sediment in French Guiana

Jennifer Hellal [1,*], Jörg Schäfer [2], Régis Vigouroux [3], Laurent Lanceleur [2] and Valérie Laperche [1]

1 BRGM, F-45060 Orléans, France; v.laperche@brgm.fr
2 University of Bordeaux, UMR CNRS 5805 EPOC, Allée Geoffroy St. Hilaire, 33615 Pessac CEDEX, France; jorg.schafer@u-bordeaux.fr (J.S.); laurent.lanceleur@univ-pau.fr (L.L.)
3 HYDRECO Guyane SARL, Laboratoire Environnement de Petit Saut, BP 823, 97 388 Kourou CEDEX, France; regis.vigouroux@hydrecolab.com
* Correspondence: j.hellal@brgm.fr

Received: 15 October 2020; Accepted: 30 October 2020; Published: 4 November 2020

Abstract: In 2006 the use of mercury (Hg) was banned for gold mining in French Guiana. However, mining of old placers could mobilize Hg accumulated in soils and sediment. This study aimed to measure the current impact of a mining concession (Boulanger site) on the Hg load in the watershed. Turbidity, Total Mercury (THg), and Monomethylmercury (MMHg) were measured in water, suspended particulate matter (SPM), river sediment and sediments from old tailing ponds along a river section of 30 km up and downstream from a mining concession in French Guiana during a dry and a rainy season. Total dissolved Hg (THg_D) concentrations varied little from up- to down-stream but were all higher (fourfold on average) during the rainy season (3.2 to 4.4 ng L^{-1}), than during the dry season and consistent with previous data known for the Amazonian area. Dissolved MMHg ($MMHg_D$) represented up to 30% of THg_D during the dry season, which is higher than previous results (typically around 2%). Mercury concentrations in sediments were highest in the vicinity of areas affected by old (before 2006) rather than new gold mining practices. Even though Hg was banned in 2006, present gold mining practices still release natural Hg and Hg inherited from older mining practices into the watershed.

Keywords: mercury; sediment; water; SPM; gold mining; French Guiana

1. Introduction

Mercury (Hg) is a toxic metal and a global pollutant, historically used in alluvial gold-mining to amalgamate and recover gold. In 2005, small-scale gold mining was estimated to contribute to more than 10% of the annual global anthropic Hg load to the atmosphere [1]. Use of Hg for gold mining was officially banned in French Guiana in 2006. However, "inherited" Hg in old gold mined areas is still present and remains a potential source of Hg to the environment. It is estimated that for 1 g of extracted gold, ~1.3 g of Hg are lost, despite its partial recycling [2,3]. In addition to Hg inputs from gold mining, Amazonian soils naturally contain high Hg concentrations due to the equatorial pedoclimatic situation (on average 10 times higher than temperate soils) [4]. During gold mining and generally during all deforesting operations [5–8], Hg adsorbed to colloids (clay, organic matter, and iron or aluminum oxides) is exported to the watershed, contaminating downstream ecosystems. In anaerobic environments (stagnant lagoons and sediment/water interface), Hg can be methylated, mainly by sulfate and iron-reducing bacteria, to form monomethylmercury (MMHg) [9,10]. MMHg is a neurotoxin which readily accumulates in the food chain up to concentrations considered dangerous for human health (around 1 mg kg^{-1} fresh weight in carnivorous fish such as Aymara, which is twice the

World Health Organization (WHO) recommendation and 10^5–10^6 times higher than surrounding water concentrations) [11]. In local Amerindian populations, whose diet traditionally consists of freshwater fish, Hg concentrations can reach values clearly exceeding World Health Organization (WHO) safety recommendations. Indeed, concentrations of more than 10 $\mu g\ g^{-1}$ hair have been measured in these populations [12].

French Guiana has known several gold rushes towards the end of the 19th century and especially at the beginning of the 20th. The gold mining techniques at that time, mainly imported from Anglo-Saxon methods developed in the Rocky Mountains, already used Hg to amalgamate gold. During these periods, gold-mined sites in French Guiana were mainly placers in small riverbeds or lateritic colluvium exploitations. The use of river barges came later, during the second half of the 20th century. It is considered that the activity at that time was much higher than presently, considering the number of placers and the volume of mineral treated. Evidently, this gold mining activity left traces of Hg contamination both on-site and downstream from the placers in the hydric network and the long-term stability of these Hg stocks is unknown. Many of the old placers are now covered by secondary rain forest, where little erosion occurs. However, these areas are potential sources of Hg pollution and need to be investigated to estimate their potential contribution to Hg dispersion in the Guianese hydric network, especially as many still contain economically relevant amounts of gold and are being re-exploited. Although 35% of the Hg used for goldmining is estimated to be released to the atmosphere, the remaining 65% are dumped on-site as metallic Hg (0) [13]. According to Laperche et al. [14], Hg concentrations in Guiana river sediments are generally close to background soil concentrations (100 ± 50 $\mu g\ kg^{-1}$) and strongly increase close to gold mining sites. Concentrations up to 12 mg kg^{-1} have been measured in the River Approuague sediments and up to 15 mg kg^{-1} along the Eddie Creek (unpublished data). Spatial distribution in sediments along French Guiana Rivers indicates that metallic Hg lost by gold miners globally remains in the rejection zones, as contamination "hot spots" [14].

At gold-mined sites, Hg in soils either occurs naturally or is left over from past gold mining in the case of reopened old sites [8]. Due to mining, this Hg is mobilized together with soil particles and is dispersed downstream in the hydric network, where it can contaminate different compartments of the ecosystem. The water rejected from legal gold mining sites is controlled by measuring water turbidity, as high water turbidity can damage aquatic ecosystems due to light extinction [15]. Guédron et al. [16] showed that exported Hg is mainly associated to the fine suspended particulate matter (SPM) and colloids, prone to be transported over longer distances than coarse particles. However, despite this knowledge and the controls, the actual quantities of Hg rejected remain unknown, whether in pristine new placers or in re-opened old gold-mining sites that potentially contain more Hg.

The objective of the present work was to evaluate and compare the amounts of Hg exported from different gold mining areas, pristine or not, to the hydric network in contrasting hydrological situations (dry and rainy season) and to identify potential changes in water quality.

2. Materials and Methods

2.1. Study Area and Sampling Strategy

The Comté-Orapu basin covers 3359 km^2. This study focuses on the sub basin of the Boulanger Creek, a tributary of the Orapu River (Figure 1), and downstream rivers including the Comté River, which joins the Orapu, forming the Oyack River. Sampling was carried out along a section of ~30 km, from the Boulanger Creek's spring, accessed along the Molokoï path, to the Oyack River, along which water and sediment samples were collected, during a dry season in September 2012 and the following rainy season in March 2013 as described below.

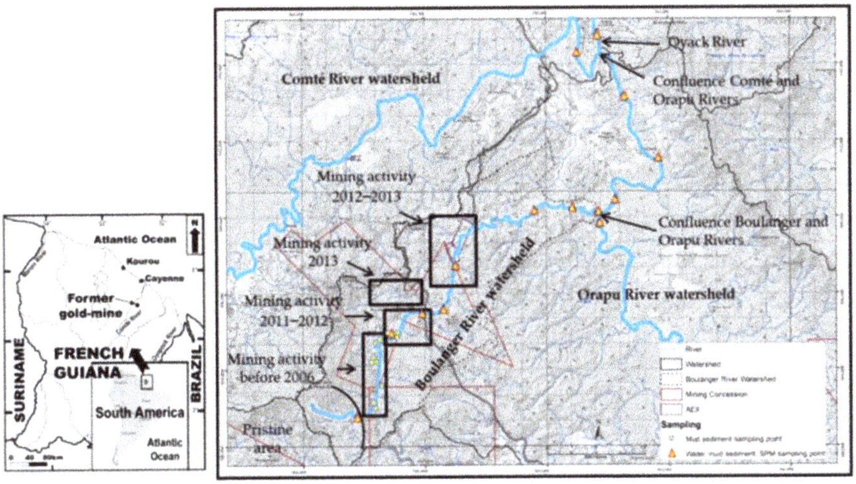

Figure 1. Map of the study area in French Guiana and the sampling points.

During each sampling campaign, 14 water (~15 cm depth) and sediment samples were collected along the Boulanger Creek (Table 1 and Figure 1) and downstream Rivers (Orapu, Comté and Oyack).

Table 1. Description of sampling points.

Water Body	Name	Description
Boulanger Creek	**B1**	Upstream Boulanger Creek
	B2	Downstream from Old Placers
	B3	Downstream 2011–2012 Placers
	B4	Fourmi Confluance
	B5	Downstream Léonce
	B6	Boulanger Creek
	B7	Boulanger Creek
	B8	Boulanger Creek
	B9	Boulanger Creek near the Confluence with the Orapu
Orapu river	**Or1**	Upstream Boulanger Confluence
	Or2	Downstream Boulanger Confluence
	Or3	Orapu Confluence Creek Marguerite
Comté	**C1**	Uptream from Orapu Confluence
Oyack	**Oy1**	Oyack Downstream from Comté and Orapu Confluence

Suspended Particulate Matter (SPM) were recovered by filtering between 250 and 500 mL water on-site through pre-weighed cellulose nitrate filters (0.45 µm; Ø47 mm; Sartorius®, Dourdan, France). The filters alone (blank) and the filters with SPM were transported to the laboratory in closed petrislides, then oven dried (50 °C, ~36 h) to constant weight. The mass of SPM sampled was determined on a precision balance (0.01 mg scales) from the difference with the initial dry filter weight. The filters were then conserved in sealed polypropylene boxes in the dark until further analysis.

Water samples were obtained by filtering 125 mL water (<0.45 µm, polyvinylidene fluoride filters (PVDF)) into pre-cleaned Teflon fluorinated ethylene propylene (FEP) bottles. Samples were acidified with suprapur HCl 0.5% *v*/*v* (Merck, Darmstadt, Germany) and stored in double polyethylene bags until analysis. All materials (PP, Teflon) in contact with samples were acid-washed (5 days in 20% HNO_3 *v*/*v* followed by 5 days in HCl 10% *v*/*v*), rinsed several times with deionized water (Milli-Q®,

Merck Millipore, Darmstadt, Germany) and dried under a laminar flow hood and stored in sealed polyethylene bags pending sampling. Polyethylene gloves were used for all handling operations.

River sediment samples were collected along the river section at the same spots as the water and SPM samples. At each sampling site, sediments were collected from just above the low water line during low tide on river banks where finer sediment tends to accumulate. It is assumed that they provide geochemical information that reflects the previous season's conditions [14]. There was no chemical (saltwater) impact of the tide on the water samples. Sediment samples were stored in polyethylene bottles and sent to the laboratory in Petit Saut (Hydreco, French Guiana, French) for treatment and analysis.

During the dry season (September 2012), sediment samples were also collected from the edges of 14 old tailing ponds (Figure 1). Some of these tailing ponds were filled more than 10 years ago and they are now covered by vegetation without having been properly rehabilitated, which makes it too dangerous to walk on the tailing pond's surface. No tailing samples were collected during the rainy season in March 2013. Indeed, as shown in the results section, Hg contents measured in the 2012 samples were similar to previous analyses carried out in 2006 [14], suggesting little evolution of the Hg concentrations in the tailing ponds.

2.2. *Mercury Analysis in Tailing Ponds and Sediment Samples*

Tailing and sediment samples were oven dried at 38 °C to constant weight and sieved (2 mm) before analysis. Total Hg (THg) concentrations in solids were determined by Atomic Absorption Spectrophotometry after dry mineralization and gold amalgamation with an Automatic Mercury Analyzer (Model AMA 254 Altec, Uherske Hradiste, Czech Republic). All particulate THg concentrations were expressed in ng g^{-1} dry sediment. Triplicates were performed for each analysis for quality assurance/quality control (QA/QC). The relative error was routinely ±5% (Relative Standard Deviation) and always under ±10%, consistently with typical relative errors reported for this method [17]. Detection limit (defined as 3 times the standard deviation (SD) of the blanks) was maximum 5 ng g^{-1}. Concentrations obtained for repeated analyses of the certified reference material (CRM) MESS-3 (Marine Sediment Reference Materials, National Research Council Canada, Ottawa, Ontario were consistently within the published concentration range (91 ± 8 ng g^{-1}).

2.3. *Dissolved Mercury (Hg_D) and Methylmercury (MMHg) Analyses*

All analytical procedures were conducted using ultra clean sample handling to avoid laboratory contamination of low-level sample extracts and a cross-contamination of high-level samples [18]. Calibration was performed by species-specific isotope-dilution by adding known amounts of ^{201}Hg-enriched Hg(II) solution (Oak Ridge National Laboratory, isotopic purity 96%) and ^{202}Hg-enriched MMHg solution ($CH_3{}^{202}HgCl$; ERM-AE670; Institute for Reference Materials and Measurements, isotopic purity of 98%) directly to the sample before the extraction procedure [19,20]. Analysis of Hg species in the organic extract was performed by Gas Chromatography (Thermo-Fischer Scientific, Focus GC, France) coupled with ICP-MS (Inductively coupled plasma mass spectrometry, Thermo-Fischer Scientific; X7, France) as previously described [21].

Detection limit (3 sigma of the blank values including reagent blank and sample handling) was 0.008 ng L^{-1}. The analytical results were continuously quality checked by analyzing an internal standard solution prepared by digesting a known amount of international certified reference material (International Atomic Energy Agency, IAEA 405) with 6 M HNO_3 [20]. Accuracy was 94 ± 4% and precision based on replicate analyses (n = 12) of the standard solution was better than 5% (RSD).

2.4. *Mercury Concentrations in Suspended Particulate Matter*

Total Hg in SPM was measured by Atomic Absorption Spectrometry with a Direct Mercury Analyzer 80 (MILESTONE, Sorisole, Italy), i.e., analytical equipment using procedures similar to those used for THg analyses in sediment (see above). Quality control was performed by measuring blanks and certified materials of different origins (estuary, sea or river sediments) and concentrations (from 30 to 1600 ng g^{-1}; NCS DC 70317 (Tibet sediment) from Laboratory Standards (Budapest, Hungary), IAEA-433 (Marine Sediment) from IAEA Reference Products for Environment and Trade, BCR-320R (Channel sediment) and BCR-280R (Lake sediment) from the Joint Research Centre (Geel, Belgium), after each set of 5 samples. Accuracy was evaluated to ~90% and precision was better than 5% (RSD).

2.5. Turbidity

Turbidity is water cloudiness caused by suspended and dissolved particles. As water turbidity is mainly caused by the presence of SPM [22], turbidity has often been used to estimate SPM concentrations. Turbidity is a function not only of SPM but also of the sizes, shapes, and composition/colour of the particles. These variables may introduce bias to SPM estimation. In this study, turbidity was assessed by the value of Nephelometric Turbidity Unit (NTU). Turbidity was measured with a microprocessor turbidity meter (Eutech Instrument Turbidimeter TN-100, Fisher Scientific, Illkirch, France) calibrated by using Formazin solutions of 0.02, 20, 100, and 800 NTU, according to the manufacturer's operating instructions.

Parallel measurements of SPM concentrations by filtration and weighing were performed (see above) to establish potential relationships between turbidity and SPM concentration under the different hydrological situations studied.

2.6. Water Flow and Hg Fluxes

Water flow was measured with an acoustic Doppler velocimeter ([ADV] 10-MHz Ocean ADV, Sontek Inc., San Diego, CA, USA), in 6 different places along the rivers: (1) upstream and (2) downstream the Boulanger Creek; (3) in the Orapu River, just upstream from its confluence with the Boulanger Creek; (4) in the Orapu River, just downstream from its confluence with the Boulanger Creek; (5) in the Comté River just before it flows into the Orapu; and (6) in the Oyack, just after its formation by the joining of the Comté and Orapu Rivers.

Fluxes were estimated for the SPM in g s^{-1}, for total particulate Hg adsorbed on the SPM (THg_P, μg s^{-1}) and for total dissolved Hg (THg_D, μg s^{-1}).

3. Results

3.1. Hg in Tailing Ponds

Total Hg concentrations in the 14 tailing pond samples ranged between 70 and 1020 ng g^{-1} dry sediment (Figure 2). The Hg concentrations in samples (n = 8) from the pond zones exploited before 2006 (from Creek St Roch to Creek Diable, points 17–24, Figure 2 covered the whole concentration range, whereas the recently exploited zone showed concentrations between 110 and 470 ng g^{-1} (n = 6). Except for point 4 (1020 ng g^{-1}), the average Hg concentration in pond sediments was 190 ng g^{-1}, which is very similar to values obtained in 2006 for the same ponds (180 ng g^{-1}, 26 samples from 10 ponds) [14].

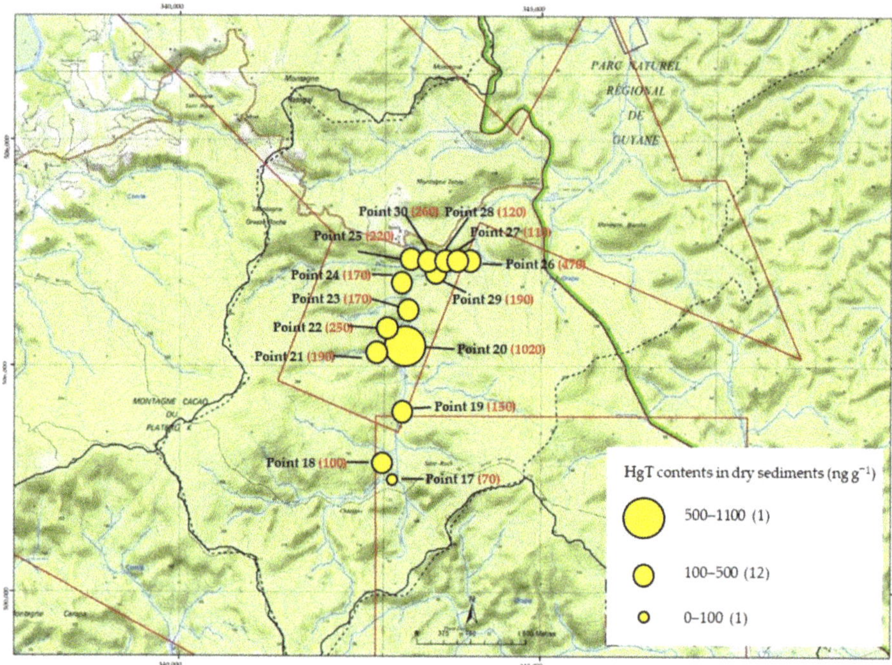

Figure 2. Localization of samples taken from old tailing ponds in the Boulanger concession in the dry season of September 2012 and total mercury (THg) contents (ng g^{-1}). Points 17–24 are tailing ponds dating before the mercury (Hg) ban for gold mining in 2006, whereas points 25–30 are more recent.

3.2. Total Hg in River Sediments and Suspended Particulate Matter in Relation to Turbidity

Total Hg in river sediments (THg) from both sampling campaigns were compared (Figure 3a). Overall, there was little difference between the dry season and the rainy season with THg concentrations varying from 40 to 380 ng g^{-1}. For both campaigns, the highest concentrations occurred at points B2 and B3 in the sector exploited prior to 2006 (B2) and in the more recently exploited area (B3).

The lowest THg in river sediment was measured at point B5, where the Boulanger Creek was deviated from its original course. The sediment at this point contained 42 ng kg^{-1}, close to the lowest value (31 ng g^{-1}) found in this area [14]. Moreover, THg concentrations at B5 were even lower in 2013 (34 ng g^{-1} dry sediment), when the creek was positioned to flow through the middle of the flats in a rehabilitated zone.

Downstream from B5, total Hg concentrations in river sediments increased regularly from 70 to maximum 150 ng g^{-1} along the Boulanger Creek to its confluence with the Orapu River (point B9) and were then constant until the Orapu and the Comté Rivers merge and form the Oyack River (points Or1 to Or3).

Total Hg concentrations in SPM collected during the dry season are comparable to those in the sediment and vary from 80 to 505 ng g^{-1} (Figure 3b). However, during the rainy season Hg concentrations in SPM increased by two- to three-fold, except in the most upstream point where values were the same (~150 ng g^{-1}) for both seasons. Maximum Hg concentrations in SPM, i.e., ~600 ng g^{-1} were higher than maximum values in sediments (~380 ng g^{-1}).

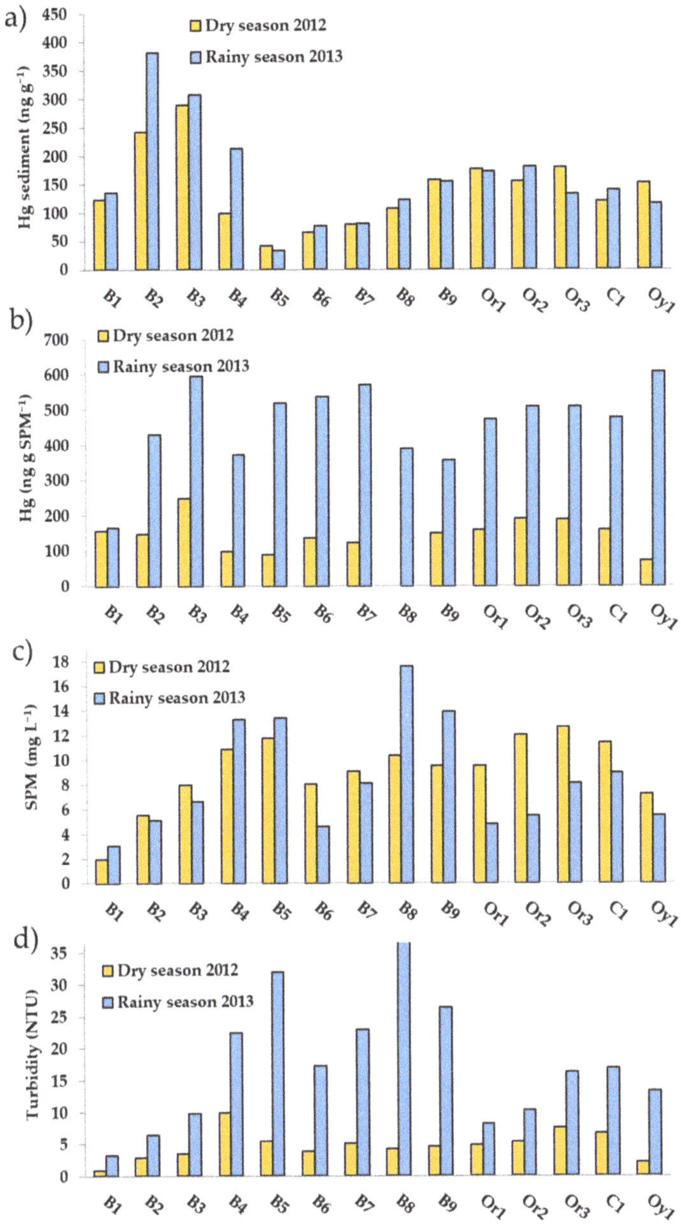

Figure 3. (**a**) Hg associated to river sediments, (**b**) Hg associated to suspended particulate matter (SPM), (**c**) suspended particulate matter (SPM) and (**d**) Turbidity at all sampling points for the dry season in September 2012 and the rainy season in March 2013. B1 to B9: Boulanger Creek, Or1 to Or3: Orapu River, C1: Comté River and Oy1: Oyack River.

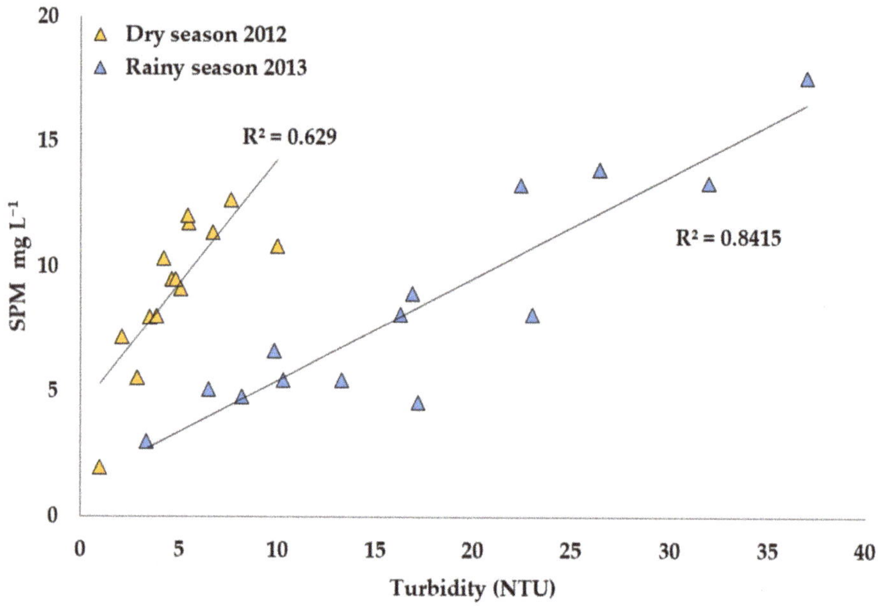

Figure 4. Relation between SPM and Nephelometric Turbidity Unit (NTU) in the dry season in September 2012 and the rainy season in March 2013.

Although Hg concentrations in SPM increased during the rainy season, the amounts of SPM in river water were globally similar for both sampling campaigns (Figure 3c) and ranged between 2 and 18 mg L^{-1}. Turbidity on the other hand was 2–9-fold higher during the rainy season with an increase in NTU along the Boulanger Creek to 38 NTU that then dropped in the Orapu and Oyak to around 10–15 NTU (Figure 3d). The lowest value (<5NTU) corresponded to an isolated spot upstream from the mining district and uninfluenced by human activities.

Accordingly, amounts of SPM did not follow the same pattern as the turbidity values. However, a linear relation was found between turbidity and SPM concentrations, that differed between the dry and rainy seasons (Figure 4).

3.3. *Hg and MMHg in River Water*

Filtered water THg_D concentrations varied little from up to downstream during the rainy season, from 3.2 to 4.4 ng L^{-1} (Figure 5) but were all higher than during the dry season (fourfold on average). During the dry season, an increase of MMHg in waters from upstream Boulanger down to the mined areas occurred. Concentrations then decreased downstream in the Orapu, probably because of the increase of the flow rate of the Orapu River (80 m^3 s^{-1}, SM-1) compared to the discharge of the Boulanger Creek (2.2 m^3 s^{-1}, SM-1). In contrast, during the rainy season, no clear pattern was observed but higher concentrations (160 pg L^{-1}) were measured at points B2, B5, B8, and C1. During the dry season, the percentage of MMHg (MMHg/THg_D) reached up to 30% downstream from the gold-mining district (Figure 5a–c), with an overall average of 13%, whereas during the rainy season it was of 3% in the same zone with a maximum of 4%.

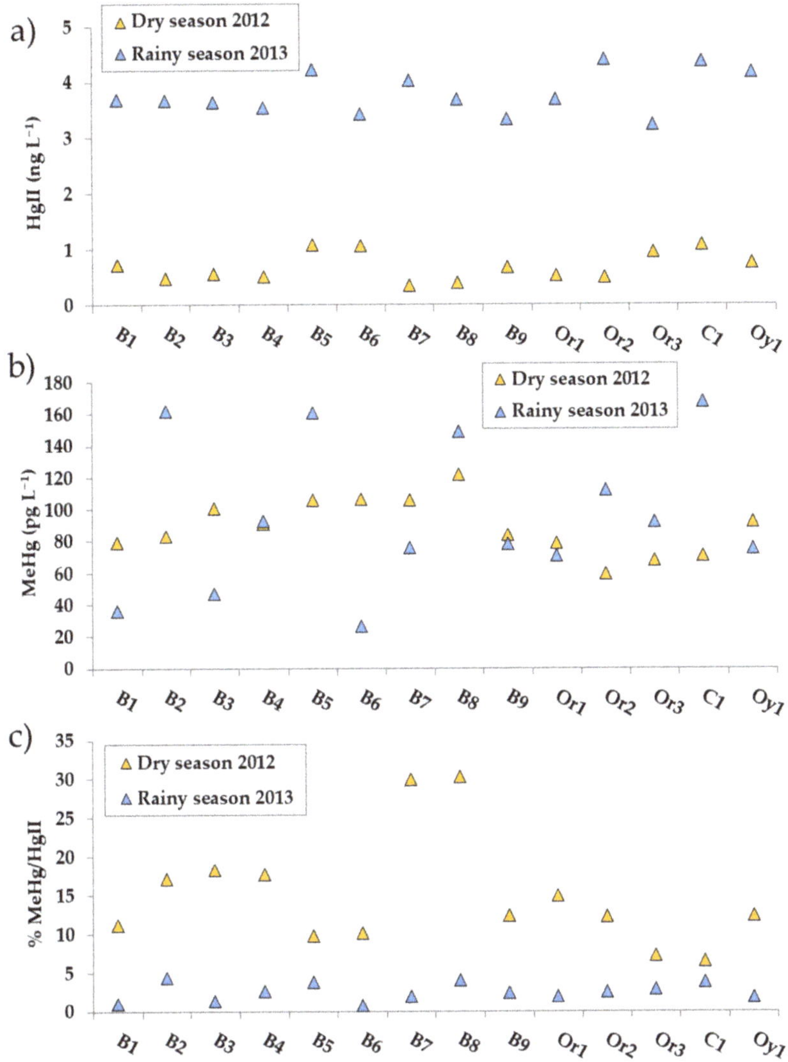

Figure 5. Dissolved Hg and monomethylmercury (MMHg) in water samples from the Oyack River catchment. (**a**): Total dissolved Hg (THg_D) (ng L^{-1}), (**b**): MMHg (pg L^{-1}) and (**c**): %MMHg/THg_D. In all three graphs, stars give data for the dry season in September 2012 and triangles for the rainy season in March 2013. B1 to B9: Boulanger Creek, Or1 to Or3: Orapu River, C1: Comté River and Oy1: Oyack River.

3.4. SPM and Hg Global Export

Based on the water discharges measured at different points during the dry and rainy seasons (SM-1), estimations of the amount of exported SPM, particulate Hg or Hg_P and THg_D were made for both seasons (Figure 6). Between the dry and the rainy season, water discharges increased by a factor of 3.4 to 4.1 at all sampling points (SM-1). There was also a similar increase (between 5.6 and 5.7) from up- to downstream along the Boulanger Creek. During the dry season, the area around the Boulanger Creek and the Comté River showed less gold mining activity than was expected. This induced a decrease in SPM export. During the dry season also, SPM exportations did not increase from upstream to downstream probably due to a balance between sedimentation and flux that maintained the SPM charge stable. On the opposite, during the rainy season, increasing river discharges resulted in higher erosion and transport, leading to an increase in SPM charge from upstream to downstream. Depending on the sampling points, the increase in Hg_P varied from 2 (upstream Orapu River) to 15 (Oyack River) between the dry and rainy season, reflecting the related soil erosion, whereas the increase in Hg_D was constant (~20 fold) whatever the sampling point.

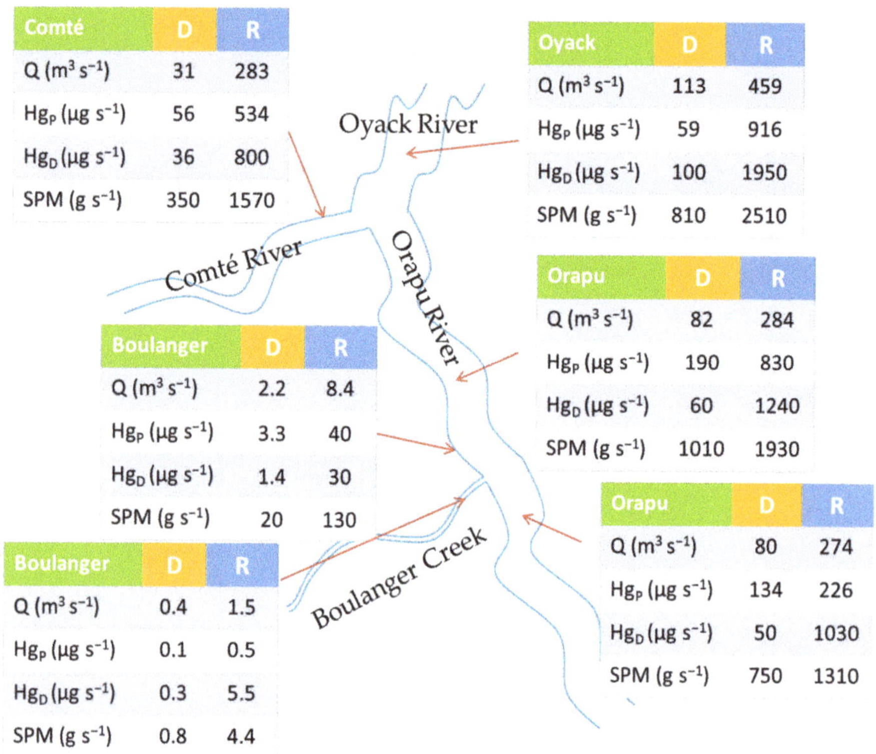

Figure 6. Comparison of water flux in the Orapu-Comté watershed and associated estimations of SPM and Hg export during the dry and rainy sampling campaigns (September 2012 and March 2013), (dry season (D); rainy season (R); particulate Hg (Hg_P); dissolved Hg (Hg_D).

4. Discussion

4.1. Hg in Tailing Ponds Remains Constant over Time

As the concentrations measured in 2012 were very similar to those measured in 2006, it appears that the Hg in the tailing ponds remains stable over time, at least at the pluri-annual time scale. The range of values found here are much lower than Hg contents reported in Venezuela [23], where Hg in pond sediments ranged from 160 ng g^{-1} to 542 µg g^{-1}, but comparable to those found in abandoned ponds in Colombia [24]. Here the authors measured THg in sediments from 27 tailing ponds, abandoned since 1997 and found between 39 and 1271 ng g^{-1} dry sediment.

Although it was expected to find higher amounts of Hg in the ponds dating from before the Hg ban for goldmining in 2006, some relatively high concentrations were also found in the more recently exploited areas. The typical background sediment concentration is around 100 ng g^{-1} sediment [14] and all the samples, except for point 17 (Figure 2), where THg was 70 ng g^{-1}, were above this value. In the zone exploited after 2006, high Hg in the tailing ponds could be a legacy from older exploitation periods.

4.2. Recent Goldmining Activity Increased THg in River Sediment Despite the Hg Ban in 2006

Overall values of THg in river sediments varied between 40 and 380 ng g^{-1} and were similar in magnitude to previous results reported for French Guiana [14]. Spadini et al., [25] found 510 ng g^{-1}, 295 ng g^{-1} and 261 ng g^{-1} dry sediment at distances of respectively 0, 35, and 50 km downstream from mining areas. Goix et al. [26], measured between 32 and 358 ng g^{-1} with significantly higher values in artisanal and small-scale gold-mining (ASGM) areas along the River Oyapock. These results are also close to values of 20 to 53 ng g^{-1} reported for river sediments along the Malinoski–Tambopata river system in Peru, a river affected by artisanal and small-scale gold-mining (ASGM) [27]. In another Peruvian river system, the Madre de Dios, also impacted by ASGM [28], authors reported Hg levels in sediments ranging from 9 ng g^{-1} upstream up to 95.3 ng g^{-1} downstream from the gold mined areas. More recently in Brazil, [29], measured between 19 and 155 ng g^{-1} along the River Tapajos. Again, highest values were observed in the vicinity of gold mining areas. However, all these Hg concentrations in goldmining-affected sediments of the Amazonian water basin were clearly lower than those reported for River Gambia in Senegal, where Hg in sediments reached 1160 +/- 800 ng g^{-1} in an area affected by ASGM [30].

In the present study, the highest concentrations of Hg in sediment occurred at points B2, B3, and B4 along the Boulanger Creek. These sites are both located directly downstream from the old tailing ponds, from where Hg-rich sediment can be exported and from the most recently exploited mining area (2011–2012), where bare soils are likely to favor soil erosion. These re-exploited old placers probably are sources of both Hg naturally present in soil and Hg leftovers from old mining practices, which would explain the increase at point B4 between the dry and rainy season (THg doubled from 100 to 200 ng g^{-1}) in relation to the onset of the exploitation next to the nearby Fourmi Creek in 2013. Previously, in a study conducted on the Tapajos River in Brazil, Telmer et al., [31] concluded that mining operations were at the origin of Hg anomalies, not from the amalgamation process itself, but from the increased erosion during the mining operations. This re-mobilizes the mercury contained naturally in soils and mercury resulting from historical mining activities that then migrates to the river. These results are coherent with our results and are supported by recent research conducted by Goix et al., [26] using an isotopic approach. These authors showed that up to 78% of Hg in sediments downstream from the gold mining site was of anthropogenic origin, of which more than half derived from the release of liquid Hg (0) and the rest from erosion of surrounding Hg-contaminated soils.

Outside the mining zones, THg concentrations in sediment were of the same order as those measured previously in this area [14]. Accordingly, the relatively low concentrations measured upstream from the mining areas (point B1) were similar to previously reported values for uncontaminated Amazonian sediments [32,33].

The decrease in the Hg load in sediments downstream from the Boulanger Creek and especially in the larger rivers such as the Orapu and the Comté is supposedly due to the huge dilution with waters draining the whole watershed and highlight the locally high values of creeks such as the Boulanger, impacted by goldmining.

4.3. Greater SPM Hg Load during the Rainy Season All along the Water Shed

Suspended Particulate Matter (SPM) increased along the Boulanger Creek, from 2 mg L^{-1} at point B1 to a maximum of 18 mg L^{-1} at B8. Downstream from there, values were between 4 and 8 mg L^{-1} during the rainy season and 8–12 mg L^{-1} during the dry season. Mining activity was low that year on the Boulanger Creek, which could explain the small difference between the two seasons, but also a natural dilution effect can reduce SPM concentrations. Due to the low erosion rates of the Precambrian Guiana Shield, French Guiana Rivers naturally have very low SPM values [34]. Amounts of SPM observed in the present work are similar to average values reported by these authors along other Guiana Rivers, i.e., the Oyapock and the Maroni Rivers.

During the rainy season, the Hg load in SPM increased by 2–3-fold at all sampling points except B1. During the dry season, concentrations were between 100 and 250 ng g^{-1} while during the rainy season they were between 400 and 600 ng g^{-1}. It would seem that the SPM mobilized during the rainy season contains more Hg than during the dry season. Roulet et al. [33,35,36] reported THg concentrations from 167 to 425 ng g^{-1} (dry weight) in SPM from clear waters (such as those in French Guiana), on a commercial sector of the Tapajos River, and from 154 to 355 ng g^{-1} (dry weight) in a zone 50 km from the mining district. The THg concentrations in the present study were slightly higher than this, but clearly lower than those reported for the Rio Négro, where THg reached values up to 2070 ng g^{-1} [32]. However, in 2006 [16], during the rainy season at point B5 (down-stream Leoncé; Figure 1), THg concentrations were 1420 ng g^{-1} (dry weight) i.e., three times higher than in the present study, which may be due to the fact that during this period there was active mining upstream from site B5. In other Amazonian watersheds, similar or higher values have been reported, e.g., between 400 and 4000 ng g^{-1} were found along the ASGM-affected Malinoski–Tambopata river system in Peru [27]. Furthermore, Hg concentration ranges of 3.2 to 133 ng g^{-1} in SPM during the dry season, and 13.3 to 42 ng g^{-1} during the rainy season were found along the Madre de Dios River [28]. In Brazil, along the Tapajos River, Hg in unfiltered water varied between 0.65 ng L^{-1} and 23.8 ng L^{-1} in tributaries closer to gold mining activities. Given that Hg is mainly transported in the particulate fraction, this maximum value corresponds to around 86 ng Hg g^{-1} SPM [29].

4.4. Relation between Turbidity and SPM Was Different According to the Season

Highest turbidity values occurred along the Boulanger Creek in areas flowing through active mining zones and downstream from these zones. Turbidity values decreased sharply in the Orapu River and then increased steadily down towards the Oyack River but not to values as high as in the Boulanger Creek. Turbidity must not be regarded as proportional to SPM concentration but as one effect of SPM among others on water physical properties such as transparency. Thus, although SPM concentrations during the rainy season are comparable to those of the dry season, turbidity can be higher. This is partly explained by the nature of SPM, which can vary from one river to another depending on the organic and mineral load and influence light reflection. Indeed, changes in turbidity per unit of SPM are likely associated with changes in particles morphology and also in changes in size as large particles scatter light less efficiently than smaller ones [37,38]. The fact that, for a similar SPM concentration, turbidity in the studied rivers was clearly higher during the rainy season, may also suggest the presence of high amounts of colloidal/dissolved components, such as humic and/or fulvic acids due to increased soil leaching [38]. Two distinct linear relationships were obtained for the rainy and dry season between SPM and turbidity, possibly linked to the different nature of SPM during the two seasons. Further work to characterize these differences could also contribute to explaining the Hg load. Turbidity could provide an easy way to control gold mining sites and might be an efficient monitor of correct use of tailing ponds, although it appears here that for a same turbidity measure, the SPM load is not the same depending on the season, and thus the amounts of transported Hg are also different.

4.5. River Water THg_D and $MMHg_D$

As in previous studies, the Hg fraction in filtered water was very low compared to the particulate fraction. Indeed, SPM remains the major pathway for river borne Hg as Hg bound to SPM [31]. Concentrations of THg_D averaged 0.25 ng L^{-1} in the dry season of 2012 and 3.70 ng L^{-1} in the rainy season of 2013. These concentrations are comparable to those found in unpolluted waters [39] and are similar to those measured previously in the Boulanger Creek in 2005–2006 [16].

During the dry season, concentrations of $MMHg_D$ were globally higher along the Boulanger Creek than further downstream, whereas during the rainy season no clear pattern was found and concentrations fluctuated between low (20 pg L^{-1}) to high concentrations (160 pg L^{-1}). In comparison, [16] found concentrations of 24–25 pg L^{-1} in the Boulanger Creek in 2005–2006. Although highest concentrations of $MMHg_D$ were measured during the rainy season, during this period it represented a very small (<3%) fraction of THg_D, whereas during the dry season this ratio went up to 30% at points B7 and B8. Such increase could be due to conditions that are more stagnant during the dry season favoring anaerobic conditions and Hg methylation.

4.6. SPM and Hg Global Export

Exported SPM from the Boulanger Creek was low compared to the Orapu River; however, it is possibly the multitude of such small creeks impacted by goldmining activities which results in the concentrations observed downstream. An approach such as the one carried out by Gallay et al. [34], who estimated the impact of land degradation on sediment fluxes in the Maroni and Oyapock River catchments, coupled to Hg measurements would yield further information on the impact of modern goldmining on Hg fluxes.

5. Conclusions

Regardless of the process used for the extraction of alluvial gold in French Guiana, the activity consists first of the resuspension of the clays and loose materials of the placers using pressurized water jets. Without carefully sized settling basins, this considerably increases river turbidity and Hg mobility. The hypothesis of the present study was that this phenomenon is more intense when older sites are exploited and that "inherited" Hg can also be remobilized. Based on the two sampling campaigns carried out in the dry season and in the rainy season downstream of the Boulanger mining sector, this study gives a first assessment of the spatial and seasonal variability of Hg exported from a re-exploited site. Results showed that, even in the absence of gold mining activity (during the dry season in September 2012, activity on the CMB concession was low), Hg was exported to the rivers downstream. Furthermore, when previously mined areas (with the use of Hg for amalgamation) are re-exploited, exports tended to be greater and were amplified during the rainy season.

The present work confirms previous observations reporting that soil erosion, whether due to deforestation for agricultural purposes, or soil excavation for goldmining is likely to contribute to Hg mobilization from soil to the water shed. Although the ban of Hg for amalgamation in 2006 put a stop to additional Hg input in legal mining exploitations, recent mining exploitations on sites where gold was mined during the last century by amalgamation imply a higher risk of re-mobilizing "inherited" mercury.

Author Contributions: Conceptualization, J.H., J.S., and V.L.; funding acquisition, J.H. and V.L.; investigation, J.H., R.V., L.L., and V.L.; project administration, V.L.; writing—original draft, J.H. and V.L.; writing—review and editing, J.S. and V.L. All authors have read and agreed to the published version of the manuscript.

Funding: This research was funded by the French Guiana Water Office and DEAL in the frame of the PRSE2 Guyane (Second programme for the promotion of Health and environment).

Acknowledgments: We thank Philippe Matheus, manager of the Boulanger Mining Company (CMB) for giving us access and authorization to collect samples from the CMB mining concession. We also thank Erwan Manach from CMB for helping with access to the sampling spots; Manuel Moisan and Bernard Joseph for their help with the fieldwork; and Cécile Bossy for her help in data acquisition.

Conflicts of Interest: The authors declare no conflict of interest. The funders had no role in the design of the study; in the collection, analyses, or interpretation of data; in the writing of the manuscript, or in the decision to publish the results.

References

1. Swain, E.B.; Jakus, P.M.; Rice, G.; Lupi, F.; Peter, A.M.; Jozef, M.P.; Alan, P.; Samuel, J.S.; Marcello, M.V. Socioeconomic Consequences of Mercury Use and Pollution. *Ambio* **2007**, *36*, 45–61. [CrossRef]
2. Picot, J.C.; Foucher, J.L.; Wagner, R. Production aurifère et mercure utilisé de l'origine à nos jours. *BRGM-Guyane Rapp.* **1993**, *R37837*, 18.
3. Pfeiffer, W.C.; Drude de Lacerda, L.; Malm, O.; Souza, C.M.M.; da Silveira, E.G.; Bastos, W.R. Mercury concentrations in inland waters of gold-mining areas in Rondônia, Brazil. *Sci. Total Environ.* **1989**, *87–88*, 233–240. [CrossRef]
4. Carmouze, J.P.; Lucotte, M.; Boudou, A. *Le Mercure en Amazonie, Rôle de l'Homme et de l'Environnement, Risques Sanitaires*; IRD Editions: Paris, France, 2002; p. 494, EAN13: 9782709914673.
5. Roulet, M.; Lucotte, M.; Canuel, R.; Farella, N.; Courcelles, M.; Guimarães, J.R.D.; Mergler, D.; Amorim, M. Increase in mercury contamination recorded in lacustrine sediments following deforestation in the central Amazon. *Chem. Geol.* **2000**, *165*, 243–266. [CrossRef]
6. Lacerda, L.D.; de Souza, M.; Ribeiro, M.G. The effects of land use change on mercury distribution in soils of Alta Floresta, Southern Amazon. *Environ. Pollut.* **2004**, *129*, 247–255. [CrossRef]
7. Almeida, M.D.; Lacerda, L.D.; Bastos, W.R.; Herrmann, J.C. Mercury loss from soils following conversion from forest to pasture in Rondonia, Western Amazon, Brazil. *Environ. Pollut.* **2005**, *137*, 179–186. [CrossRef] [PubMed]
8. Grimaldi, M.; Guédron, S.; Grimaldi, C. Impact of gold mining on mercury contamination and soil degradation in Amazonian ecosystems of French Guiana. In *Land-Use Change Impacts on Soil Processes: Tropical and Savannah Ecosystems*; Brearley, F.Q., Thomas, A.D., Eds.; CAB International: Wallingford, UK, 2015; pp. 95–107.
9. Compeau, G.C.; Bartha, R. Sulfate-Reducing Bacteria: Principal Methylators of Mercury in Anoxic Estuarine Sediment. *Appl. Environ. Microbiol.* **1985**, *50*, 498–502. [CrossRef] [PubMed]
10. Fleming, E.J.; Mack, E.E.; Green, P.G.; Nelson, D.C. Mercury Methylation from Unexpected Sources: Molybdate-Inhibited Freshwater Sediments and an Iron-Reducing Bacterium. *J. Appl. Environ. Microbiol.* **2006**, *72*, 457–464. [CrossRef]
11. Charlet, L.; Boudou, A. Cet or qui file un mauvais mercure. *Recherche* **2002**, *359*, 52–59.
12. Cordier, S.; Grasmick, C.; Pasquier-Passelaigue, M.; Mandereau, L.; Weber, J.P. Imprégnation de la population guyanaise par le mercure: Niveaux et sources d'exposition. *Bull. Epidémiol. Hebd.* **1997**, *14*, 59–61.
13. Telmer, K.H.; Veiga, M.M. World emissions of mercury from artisanal and small scale gold mining. In *Mercury Fate and Transport in the Global Atmosphere: Emissions, Measurements and Models*; Mason, R., Pirrone, N., Eds.; Springer: Boston, MA, USA, 2009; pp. 131–172.
14. Laperche, V.; Hellal, J.; Maury-Brachet, R.; Joseph, B.; Laporte, P.; Breeze, D.; Blanchard, F. Regional distribution of mercury in sediments of the main rivers of French Guiana (Amazonian basin). *SpringerPlus* **2014**, *3*, 1–11. [CrossRef]
15. Davies-Colley, R.J.; Smith, D.G. Turbidity Suspended Sediment, and water clarity: A Review. *JAWRA J. Am. Water Resour. Assoc.* **2001**, *37*, 1085–1101. [CrossRef]
16. Guedron, S.; Grimaldi, M.; Grimaldi, C.; Cossa, D.; Tisserand, D.; Charlet, L. Amazonian former gold mined soils as a source of methylmercury: Evidence from a small scale watershed in French Guiana. *Water Res.* **2011**, *45*, 2659–2669. [CrossRef]
17. Roos-Barraclough, F.; Givelet, N.; Martinez-Cortizas, A.; Goodsite, M.E.; Biester, H.; Shotyk, W. An analytical protocol for the determination of total mercury concentrations in solid peat samples. *Sci. Total Environ.* **2002**, *292*, 129–139. [CrossRef]
18. Cossa, D.; Gobeil, C. Mercury speciation in the Lower St. Lawrence Estuary. *Can. J. Fish. Aquat. Sci.* **2000**, *57*, 138–147. [CrossRef]
19. Monperrus, M.; Krupp, E.; Amouroux, D.; Donard, O.F.X.; Martín-Doimeadios, R.C.R. Potential and limits of speciated isotope-dilution analysis for metrology and assessing environmental reactivity. *TrAC Trends Anal. Chem.* **2004**, *23*, 261–272. [CrossRef]
20. Schäfer, J.; Castelle, S.; Blanc, G.; Dabrin, A.; Masson, M.; Lanceleur, L.; Bossy, C. Mercury methylation in the sediments of a macrotidal estuary (Gironde Estuary, south-west France). *Estuar. Coast. Shelf Sci.* **2010**, *90*, 80–92. [CrossRef]

21. Martín-Doimeadios, R.C.R.; Krupp, E.; Amouroux, D.; Donard, O.F.X. Application of Isotopically Labeled Methylmercury for Isotope Dilution Analysis of Biological Samples Using Gas Chromatography/ICPMS. *Anal. Chem.* **2002**, *74*, 2505–2512. [CrossRef]
22. Ives, K.J.; Atkin, J.R.; Thompson, R.P. Measurement of turbidity. *Effl. Water Treat. J.* **1968**, *3*, 342–348.
23. Santos-Francés, F.; García-Sánchez, A.; Alonso-Rojo, P.; Contreras, F.; Adams, M. Distribution and mobility of mercury in soils of a gold mining region, Cuyuni river basin, Venezuela. *J. Environ. Manag.* **2011**, *92*, 1268–1276. [CrossRef]
24. Gutiérrez-Mosquera, H.; Marrugo-Negrete, J.; Díez, S.; Morales-Mira, G.; Montoya-Jaramillo, L.J.; Jonathan, M.P. Distribution of chemical forms of mercury in sediments from abandoned ponds created during former gold mining operations in Colombia. *Chemosphere* **2020**, *258*, 127319. [CrossRef]
25. Spadini, L.; Charlet, L. Distribution of anthropogenic mercury in French Guyana river sediments downstream from gold mining sites. *J. Phys. IV Fr.* **2003**, *107*, 1263–1266. [CrossRef]
26. Goix, S.; Maurice, L.; Laffont, L.; Rinaldo, R.; Lagane, C.; Chmeleff, J.; Menges, J.; Heimbürger, L.-E.; Maury-Brachet, R.; Sonke, J.E. Quantifying the impacts of artisanal gold mining on a tropical river system using mercury isotopes. *Chemosphere* **2019**, *219*, 684–694. [CrossRef]
27. Moreno-Brush, M.; Rydberg, J.; Gamboa, N.; Storch, I.; Biester, H. Is mercury from small-scale gold mining prevalent in the southeastern Peruvian Amazon? *Environ. Pollut.* **2016**, *218*, 150–159. [CrossRef] [PubMed]
28. Diringer, S.E. River transport of mercury from artisanal and small-scale gold mining and risks for dietary mercury exposure in Madre de Dios, Peru. *Environ. Sci.* **2015**, *17*, 478–487. [CrossRef]
29. Lino, A.S.; Kasper, D.; Guida, Y.S.; Thomaz, J.R.; Malm, O. Total and methyl mercury distribution in water, sediment, plankton and fish along the Tapajós River basin in the Brazilian Amazon. *Chemosphere* **2019**, *235*, 690–700. [CrossRef] [PubMed]
30. Niane, B.; Guédron, S.; Feder, F.; Legros, S.; Ngom, P.M.; Moritz, R. Impact of recent artisanal small-scale gold mining in Senegal: Mercury and methylmercury contamination of terrestrial and aquatic ecosystems. *Sci. Total Environ.* **2019**, *669*, 185–193. [CrossRef]
31. Telmer, K.; Costa, M.; Angélica, R.S.; Araujo, E.S.; Maurice, Y. The source and fate of sediment and mercury in the Tapajós River, Pará, Brazilian Amazon: Ground- and space-based evidence. *J. Environ. Manag.* **2006**, *81*, 101–113. [CrossRef] [PubMed]
32. Maurice-Bourgoin, L.; Quemerais, B.; Moreira-Turcq, P.; Seyler, P. Transport, distribution and speciation of mercury in the Amazon River at the confluence of black and white waters of the Negro and Solimões Rivers. *Hydrol. Process.* **2003**, *17*, 1405–1417. [CrossRef]
33. Roulet, M.; Lucotte, M.; Canuel, R.; Rheault, I.; Tran, S.; Gog, Y.G.D.F.; Farella, N.; do Vale, R.S.; Passos, C.J.S.; da Silva, E.D.J.; et al. Distribution and partition of total mercury in waters of the Tapajós River Basin, Brazilian Amazon. *Sci. Total Environ.* **1998**, *213*, 203–211. [CrossRef]
34. Gallay, M.; Martinez, J.-M.; Allo, S.; Mora, A.; Cochonneau, G.; Gardel, A.; Doudou, J.-C.; Sarrazin, M.; Chow-Toun, F.; Laraque, A. Impact of land degradation from mining activities on the sediment fluxes in two large rivers of French Guiana. *L. Degrad. Dev.* **2018**, *29*, 4323–4336. [CrossRef]
35. Roulet, M.; Lucotte, M.; Saint-Aubin, A.; Tran, S.; Rhéault, I.; Farella, N.; Da Silva, E.D.J.; Dezencourt, J.; Passos, C.J.S.; Soares, G.S.; et al. The geochemistery of mercury in central Amazonian soils developed on the Alter-do-Chao formation of the lower Tapajos River Vally, Para state, Brazil. *J. Sci. Total Environ.* **1998**, *223*, 1–24. [CrossRef]
36. Roulet, M.; Lucotte, M.; Canuel, R.; Farella, N.; Goch, Y.G.D.F.; Peleja, J.R.P.; Guimarães, J.R.D.; Mergler, D.; Amorim, M. Spatio-temporal geochemistry of mercury in waters of the Tapajós and Amazon rivers, Brazil. *Limno. Oceanogr.* **2001**, *46*, 1141–1157. [CrossRef]
37. Kirk, J.T.O. *Light and Photosynthesis in Aquatic Ecosystems*; Cambridge University Press: Cambridge, UK, 1994; p. 524. [CrossRef]

38. Pfannkuche, J.; Schmidt, A. Determination of suspended particulate matter concentration from turbidity measurements: Particle size effects and calibration procedures. *Hydrol. Process.* **2003**, *17*, 1951–1963. [CrossRef]
39. Coquery, M.; Cossa, D.; Azemard, S.; Peretyazhko, T.; Charlet, L. Methylmercury formation in the anoxic waters of the Petit-Saut reservoir (French Guiana) and its spreading in the adjacent Sinnamary River. *J. Phys. IV Fr.* **2003**, *107*, 327–331. [CrossRef]

applied sciences

MDPI

Article

Accumulation of Methylmercury in the High-Altitude Lake Uru Uru (3686 m a.s.l, Bolivia) Controlled by Sediment Efflux and Photodegradation

Stéphane Guédron [1,2,*], Dario Achá [3,*], Sylvain Bouchet [4], David Point [3,5], Emmanuel Tessier [4], Carlos Heredia [3], Stéfany Rocha-Lupa [3], Pablo Fernandez-Saavedra [3], Marizol Flores [2,5], Sarah Bureau [1], Israel Quino-Lima [2] and David Amouroux [4,*]

1 Univ. Grenoble Alpes, Univ. Savoie Mont Blanc, CNRS, IRD, IFSTTAR, ISTerre, 38000 Grenoble, France; sarah.bureau@univ-grenoble-alpes.fr

2 Laboratorio de Hidroquímica—Instituto de Investigaciones Químicas—Universidad Mayor de San Andrés, Campus Universitario de Cota-Cota, casilla, 3161 La Paz, Bolivia; marizol_775@hotmail.com (M.F.); israelquino@hotmail.com (I.Q.-L.)

3 Unidad de Calidad Ambiental (UCA)—Instituto de Ecología—Universidad Mayor de San Andrés, Campus Universitario de Cota Cota, casilla, 3161 La Paz, Bolivia; david.point@ird.fr (D.P.); cr_heredia_a@hotmail.com (C.H.); bio_fany@hotmail.com (S.R.-L.); pablo86fernandez@hotmail.com (P.F.-S.)

4 Univ. de Pau et des Pays de l'Adour, E2S UPPA, CNRS, IPREM, Institut des Sciences Analytiques et de Physico-chimie pour l'Environnement et les Matériaux, Pau, France; sylvain.bouchet@eawag.ch (S.B.); emmanuel.tessier@univ-pau.fr (E.T.)

5 Géosciences Environnement Toulouse, UMR5563—IRD UR 234, Université Paul Sabatier, 14 Avenue Edouard Belin, 31400 Toulouse, France

* Correspondence: stephane.guedron@ird.fr (S.G.); darioAchá@yahoo.ca (D.A.); david.amouroux@univ-pau.fr (D.A.); Tel.: +33-476-63-59-28 (S.G.)

Received: 14 October 2020; Accepted: 4 November 2020; Published: 9 November 2020

Abstract: In shallow aquatic environments, sediment is a significant source of monomethylmercury (MMHg) for surface water (SW). High-altitude aquatic ecosystems are characterized by extreme hydro-climatic constraints (e.g., low oxygen and high UV radiation). We studied, during two seasons, the diel cycles of MMHg in SW and sediment porewaters (PW) of Lake Uru Uru (3686 m a.s.l, Bolivia) contaminated by urban and mining activities. Our results show that diel changes in SW MMHg concentrations (up to 1.8 ng L^{-1}) overwhelm seasonal ones, with higher MMHg accumulation during the night-time and the dry season. The calculation of MMHg diffusive fluxes demonstrates that the sediment compartment was the primary source of MMHg to the SW. Most MMHg efflux occurred during the dry season (35.7 ± 17.4 ng m^{-2} day^{-1}), when the lake was relatively shallow, more eutrophicated, and with the redoxcline located above the sediment–water interface (SWI). Changes in MMHg accumulation in the PWs were attributed to diel redox oscillations around the SWI driving both the bacterial sulfate reduction and bio-methylation. Finally, we highlight that although MMHg loading from the PW to the SW is large, MMHg photodegradation and demethylation by microorganisms control the net MMHg accumulation in the water column.

Keywords: monomethylmercury; water–sediment interface; diel and seasonal cycles; photodegradation

1. Introduction

Lake sediment is known to be an important source of mercury (Hg) species, including the neurotoxic monomethylmercury (MMHg) for overlying waters [1], through their diffusion and advection under dissolved and colloidal phases [2–6]. Hg methylation mainly occurs in sub to anoxic surficial sediments or biofilms developed at the sediment–water interface (SWI), implying various microorganisms such as methanogens, sulfate-reducing (SRB), and iron-reducing (IRB) bacteria [7–10]. In shallow environments (e.g., shallow lakes, lagoons, ponds), both the shear stress and the bio-irrigation driven by tides, wind, and benthic organisms enhance the upward release of elements contained in the sediment porewater towards the water column [2–6,11–13]. Such MMHg exchange is mainly regulated by the position of the redox front around the SWI where it is mainly produced, the sediment structure (e.g., porosity, tortuosity), and the abundance and type of (in)organic ligands [14,15] and colloids [16].

In high-altitude shallow lakes, extreme hydro-climatic constraints (e.g., high UV radiation and low oxygen) result in large diel variations of the water column's biogeochemical and physical parameters that can overwhelm seasonal amplitude [17]. This is the case in Lake Uru Uru (3686 m above sea level (a.s.l.)), one of the most Hg-contaminated lakes of the Bolivian Altiplano [18,19]. Previous studies [18,19] have shown that urban and mining discharges to the lake, combined with the strong diurnal biogeochemical gradients, have resulted in large diel variability in MMHg levels (0.2 to 4.5 ng L^{-1}) in the water column. The quantification of methylation potential during incubation experiments with enriched isotopes has suggested that sediments could be the major source of MMHg for the water column [18]. Nevertheless, the sediment–water exchange has not yet been assessed.

In this work, we studied Hg and MMHg sediment–water column exchanges in Lake Uru Uru during two 36-h cycles performed at the end of the wet (May) and at the end of the dry (November) seasons of 2014. Hg speciation was measured in the water column (i.e., filtered, particulate, and dissolved gaseous Hg) and in the surficial sediment porewater. We also calculated Hg and MMHg diffusive fluxes at the SWI to assess the sediment contribution in MMHg effluxes toward the water column. The final objective was to identify the biogeochemical and physicochemical factors influencing these sediment–water exchanges and to evaluate those regulating the accumulation of MMHg in the water column.

2. Materials and Methods

2.1. Lake Uru Uru General Settings

Lake Uru Uru (3686 m a.s.l.) is located in the central part of the Bolivian Altiplano region (Figure 1), south of Oruro city. This shallow aquatic ecosystem is a human-made reservoir mainly supplied by the Rio Desaguadero waters. Depending on the season, the lake surface and depth vary between 120 and 350 km^2 and from 0.25 to 1 m, respectively, representing a change in lake volume up to 12 times between the two seasons.

Historical and current mining (e.g., Au, Ag, and Sn) and smelting activities in the region have led to the contamination of the lake by various metals, including Hg [19,20]. In addition, the recent urban development has resulted in the almost permanent eutrophication of this shallow aquatic ecosystem [21]. A sharp contrast exists between the northern and southern parts of the lake, with a higher density of sedge (*Schoenoplectus totora*), grass (*Ruppia* spp.), and algae (Characeae) in the southern part [22]. Previous investigations demonstrated that the diel variability (e.g., temperature, oxygen) in the lake could overwhelm seasonal variations [18]. Furthermore, the synergistic effect of mining (i.e., acid mine drainage (AMD)) and urban effluents from both the Oruro and Huanuni areas (Figure 1) to this shallow lake were proven to enhance Hg methylation, with the sediment being the main source of MMHg during the dry season [19].

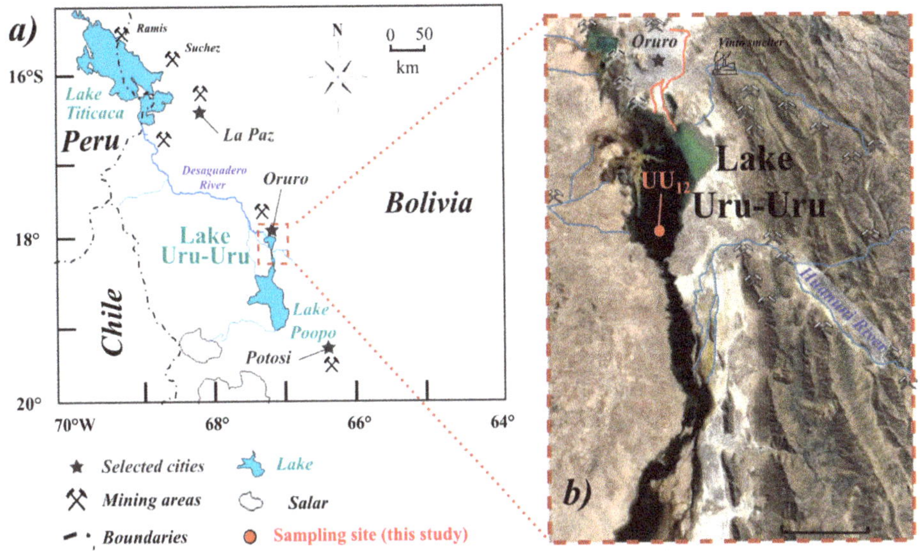

Figure 1. (**a**) Study site location: location of Lake Uru Uru in the Bolivian Altiplano and (**b**) sampling site (UU_{12}) in the southern part of the lake.

2.2. Sampling Location, Strategy, and Elemental Analysis

Surface water and sediment porewater samples were collected during two campaigns performed in May (end of the wet season) and November (end of the dry season) 2014. Surface water was sampled with a Teflon-coated Go-Flo trace metals sampler. Samples were then stored unfiltered or filtered (0.22 µm PVDF) in acid-washed FEP Teflon containers and acidified (except for dissolved gaseous mercury samples) with HCl (0.5%, *v*/*v*, Ultrex grade—Baker). Surface sediment porewaters (0–5 cm) were sampled with 5 cm long microporous polymer tube samplers (RhizonCSS®, 0.2 µm porosity; Rhizosphere Research Products, the Netherlands) over ~4 h, following published protocols [3,23].

Dissolved gaseous mercury (DGM) was purged from unfiltered water samples on-site within ~0.5 h after sampling and collected on gold-coated sand traps [24]. DGM was analyzed by thermodesorption (600 °C) and cold vapor atomic fluorescence spectrometry (CV-AFS) [24]. Filtered (F) and unfiltered (UNF) total mercury concentrations (THg) in surface and pore waters were determined by cold vapor atomic fluorescence spectrometry (CV-AFS) after conversion of all mercury species into Hg0 followed by detection using a Tekran® (Model 2500, Canada) [25,26]. Filtered monomethylmercury ($MMHg_F$) concentrations were analyzed using an ethylation purge and trap-gas chromatograph-AFS analyzer (MERX System, Brooks Rand®, Seattle, WA, USA) following published protocols [27,28]. All samples were run in duplicate and quantified using the standard addition technique. Unfiltered monomethylmercury ($MMHg_{UNF}$) concentrations were analyzed by propylation ID-GC–ICPMS (S.I.2.2) following published protocols [4,29,30].

Dissolved organic carbon concentrations (DOC) were determined using a Non-Dispersive Infra-Red (NDIR) CO_2 Shimadzu® (Model VCSN, Japan) spectrometer after humid oxidation in a sodium persulfate solution at 100 °C. Major elements (Ca, Si, K, Na, Mg, and S) in filtered water (SW and PW) were measured by ICP-AES (Varian model 720ES) within the analytical chemistry platform of ISTerre (OSUG-France) and trace elements (Fe, Al, Mn, V, Sb) by ICP-MS (Nexion, Perkin Elmer) within the analytical chemistry platform of Univ. de Pau et des Pays de l'Adour (UPPA). Anions were measured by ionic chromatography using a 332 Mctrohm apparatus and its accuracy was evaluated with a Carl Roth 2668.1 standard. Hydrogen sulfide (H_2S) was determined using a modification of the diamine method described elsewhere [31,32] and analyzed with HPLC (Agilent) after calibration with a Radiello® solution (Code 171, Italy) for H_2S [28].

A submersible multiparameter probe (HI-9828, Hanna Instruments, Woonsocket, RI, USA) was used to characterize the water physico-chemistry (pH, conductivity, dissolved oxygen concentration and saturation, and temperature). Data collection was conducted every three seconds for two minutes at the surface (0–10 cm) and above the sediment–water interface (SWI) at each time of sampling [28].

2.3. Sediment–Water Flux Measurement and Calculation

Molecular diffusion from sediment to overlying water was estimated through the calculation of diffusive flux for Hg(II) (= THg − MMHg) and MMHg at the SWI using the measured concentration gradient and Fick's first law [33,34]:

$$Jsed = -Ø/Dsed\ (\partial C/\partial z)$$

where Jsed (ng.m^{-2} d^{-1}) is the diffusive flux from the sediment, Ø is the porosity (Ø = 0.78 [20]), Dsed is the diffusive coefficient in sediment (2×10^{-6} cm^2 s^{-1} and 1.2×10^{-5} cm^2 s^{-1} for Hg(II) and MMHg ionic/molecular diffusion coefficient in water [35]), and $\partial C/\partial z$ is the linear concentration gradient through the SWI. Positive Jsed indicates an upward-directed flux (efflux from the sediment into the overlying water column), and negative Jsed indicates a downward-directed flux (influx from the water column into the sediment).

3. Results

3.1. Diel and Seasonal Variations of Mercury Species in Surface Water

The water column was oxic during both seasons, with the lowest oxygen levels (O_2 = 76 ± 23%$_{saturation}$) and redox potential values [i.e., oxydo-reduction potential (ORP) = −8 to −72 mV] found during the dry season compared to the wet one (O_2 = 93 ± 17%$_{sat}$), where redox (i.e., ORP = 192 to −115 mV) was only negative during the night, resulting from algal respiration and sulfate reduction (H_2S = 51 ± 42 μg L^{-1}, Figure 2a,b).

Dissolved gaseous mercury (DGM = 11.2 ± 9.9 pg L^{-1}) in surface water (SW) was mainly in the form of Hg(0) [19] and was overall low and never exceeded 1% of THg_{UNF} for both seasons (Figure 2a), similar to previous studies in Lake Uru Uru [19] and Lake Titicaca [24] but in the lowest range of values reported for temperate freshwater lakes [36,37]. Higher DGM levels were found during the dry season (21.0 ± 7.5 pg L^{-1}), which were around four times higher than during the wet one (5.6 ± 5.5 pg L^{-1}).

Average unfiltered (UNF) concentrations of THg (3.3 ± 1.0 ng L^{-1}) and MMHg (0.6 ± 0.2 ng L^{-1}) were 3 to 10 times higher than those measured in Lake Titicaca [2,24] reflecting the local mining contamination and high MMHg production in this ecosystem [18,19]. Both the average filtered (F) and unfiltered (UNF) THg and MMHg levels in surface water (Figure 2c,d) were similar between the two seasons. In contrast, filtered alkali and alkaline-earth metal concentrations (e.g., Li, Na, K, Mg, Ca, and Sr) and DOC increased by two to three times during the dry season due to the evaporation and concentration of ions in water (Figure 2b and Figure S1).

The diel variability of F and UNF MMHg concentrations overwhelmed the seasonal one, with the highest values measured during the night-time when oxygen levels dropped (Figure 2a), resulting from the aquatic ecotopes' respiration. MMHg/THg percentages were high in surface waters for both seasons. However, they presented larger diel variations, with lower average values during the dry season (i.e., 35 ± 20% and 26 ± 7% for F and UNF, respectively) compared to the wet one (i.e., 42 ± 4% and 35 ± 6%, Figure 2e).

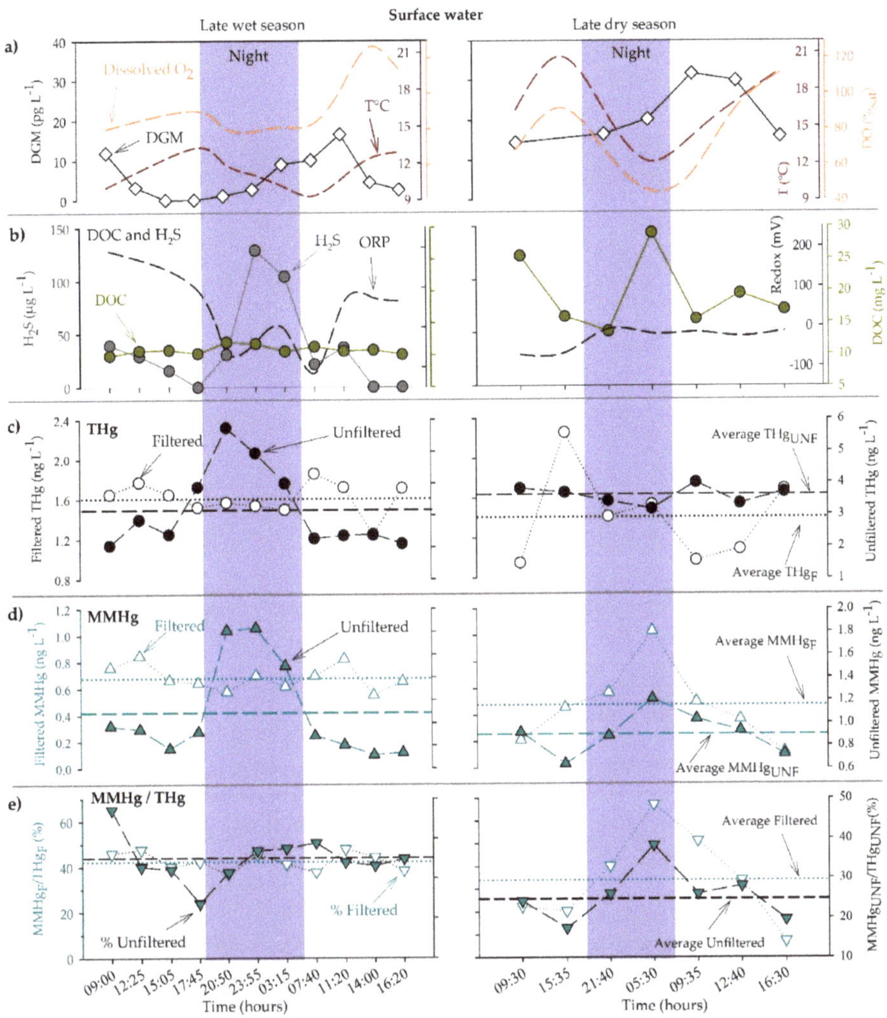

Figure 2. (**a**) Dissolved gaseous Hg (DGM), dissolved oxygen saturation percentage ($\%_{sat}$), and temperature (T °C); (**b**) dissolved organic carbon (DOC), hydrogen sulfide (H_2S), and redox potential (ORP); (**c**) filtered (F, empty symbols) and unfiltered (UNF, black symbols) THg concentrations; (**d**) filtered (F, empty symbols) and unfiltered (UNF, dark cyan symbols) MMHg concentrations; and (**e**) percentage MMHg over the THg concentrations in the F (empty symbols) and UNF (dark cyan symbols) in surface water during the wet (left panels, 20–21 May 2014) and the dry (right panel, 18th to 19th November 2014) seasons. Lines represent average concentrations for F (dotted line) and UNF (dashed line) species and ratios.

3.2. Diel and Seasonal Variations of Mercury Species in Surface Sediment Porewater

Filtered THg (8.8 ± 3.3 ng L^{-1}) and MMHg (1.7 ± 1.1 ng L^{-1}) levels in surface sediment PW (Figure 3a) were 2 to 8 times higher than those of SW. In contrast to SW, large seasonal differences were observed in the PW, with higher MMHg concentrations during the dry season (3.0 ± 1.2 ng L^{-1}) compared to the wet one (1.0 ± 0.4 ng L^{-1}, Figure 3b). Percentage MMHg/THg (Figure 3b) was also higher during the dry season (28.1 ± 11.0%) compared to the wet one (15.5 ± 8.8%), highlighting a higher accumulation of MMHg in surface sediment PW during the dry season.

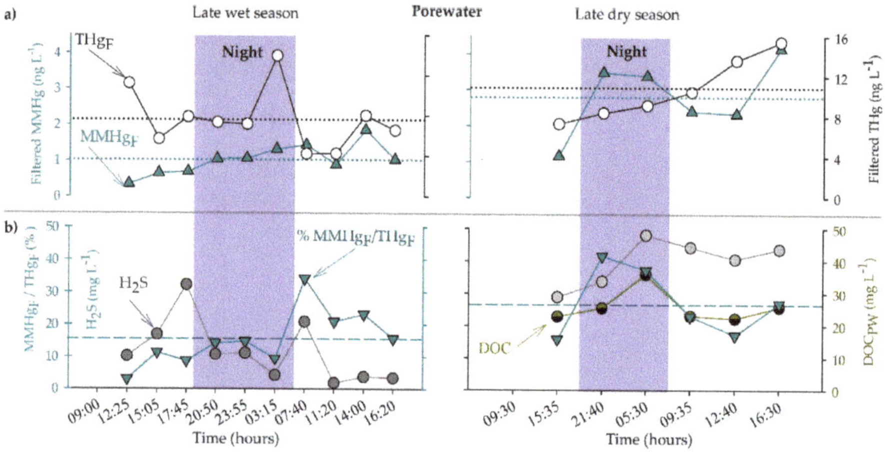

Figure 3. (**a**) Filtered (F) THg (empty circles) and MMHg concentrations (dark cyan triangles up) and (**b**) percentage MMHg over the THg concentrations (dark cyan triangle down), dissolved organic carbon (DOC, semi-filled dark yellow circles), and hydrogen sulfide (H_2S, grey circles) concentrations in sediment porewater during the wet (left panels, 20–21 May 2014) and the dry (right panel, 18th to 19th November 2014) seasons. Lines represent average concentrations for F (dotted line) and MMHg/THg (dashed line) concentrations and ratios.

Diel variations were large for THg, but no specific trend was identified except for a rising trend in concentrations during the day for the dry season. In contrast, MMHg exhibited similar trends as for SW, with rising concentration during the night for the dry season, while they did not vary significantly during the wet season.

DOC concentrations in PWs (26 ± 5 mg L^{-1}, only measured during the dry season) were, on average, 1.5 times higher than those of SW but exhibited a similar diel trend, suggesting that equilibrium was established between these two compartments. Sulfide concentrations were an order of magnitude higher than the one in SW, with at least three times higher average concentrations during the dry season (39.9 ± 7.2 mg L^{-1}) compared to the wet one (12.1 ± 9.0 mg L^{-1}, Figure 3b).

4. Discussions

4.1. DOC and Sulfides Drive MMHg Accumulation in Sediment PW and Effluxes Towards the SW

The high MMHg concentrations found in PW are consistent with a previous study in Lake Uru Uru, which showed evidence that methylation was mediated mostly by SRB in the absence of sunlight [18]. Such nutrient-rich, alkaline waters with elevated levels in DOC and sulfates were also reported to favor the activities of anaerobic bacterial communities [4,23,24,38]. During the wet season, the gradual rise in MMHg concentration in PW during the night illustrates the progressive installation of anoxia in surface sediment PW due to the rising respiration of the aquatic ecotopes, which is also reflected by the drop in dissolved oxygen in overlying waters (Figure 4a). This pattern is exacerbated during the dry season as a result of more eutrophic conditions, which enhance MMHg production and accumulation in the sediments [39]. During this latter season, both DOC and sulfide concentrations were positively correlated (R = 0.56, $p < 0.05$), supporting the notion that active sulfate reduction in PW favors DOC release in PWs during the reduction of Fe-oxy(hydr)oxides [40].

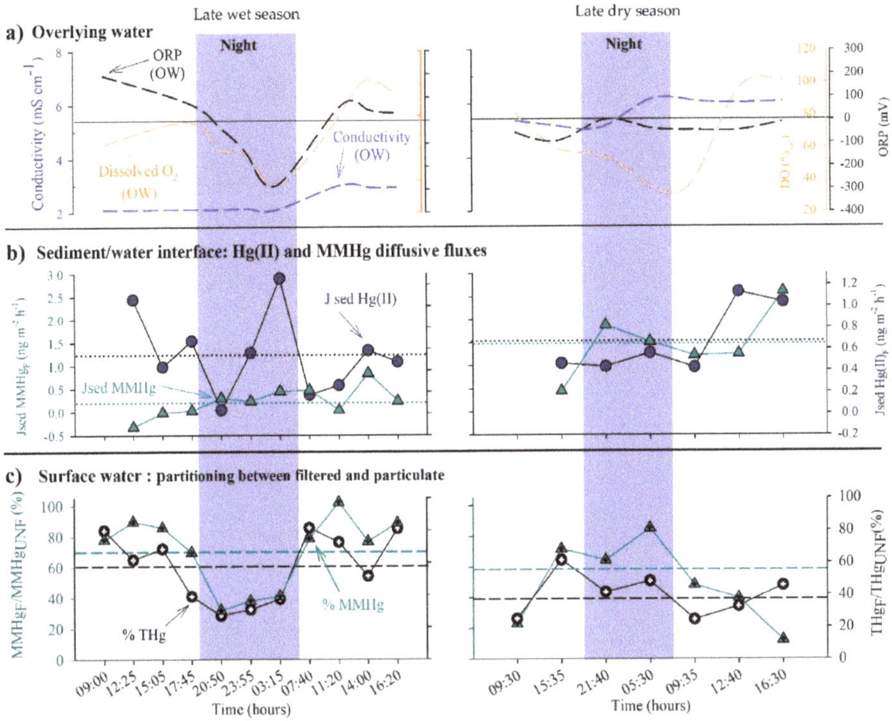

Figure 4. (**a**) Redox potential (ORP), dissolved oxygen saturation percentage (%$_{sat}$), and conductivity in overlying waters (OW) above the sediment–water interface (SWI); (**b**) filtered Hg(II) (=THg − MMHg) and MMHg diffusive fluxes (Jsed) at the SWI and (**c**) partitioning of THg and MMHg between filtered and particulate fractions in surface water, during the wet (left panels) and the dry (right panels) seasons. Lines represent averages for Jsed (dotted line) and for MMHg/THg ratios (dashed line).

Interestingly, positive correlations were also found between sulfide and MMHg concentration (R = 0.73, $p < 0.01$, Figure S3) or percentages MMHg/THg (R = 0.66, $p < 0.01$), and between percentages MMHg/THg and DOC (R = 0.66, $p < 0.01$). This suggests that filtered MMHg was preferentially bound to sulfurized dissolved organic matter (DOM). In addition, the positive correlation found between sulfide and THg concentration (Figure S2) supports a limited precipitation of inorganic Hg with free sulfide or with FeS, which would have restricted the availability of Hg(II) for bio-methylation [41–43]. Hence, this rather supports the presence of Hg-SR complexes, which were reported to increase the bioavailability of Hg(II) for bio-methylation [14,15]. In contrast, sulfide was negatively correlated with chalcophile redox-sensitive elements (i.e., Fe and V, $p < 0.01$, respectively, Figures S2 and S3), supporting the notion that the sulfate reduction in anoxic PWs favors their precipitation as authigenic sulfide minerals.

The calculation of diffusive fluxes (Jsed, Figure 4b) for MMHg and HgII (=THg − MMHg) confirms both the seasonal and diel patterns observed in the PW and SW (Figures 2 and 3). Daily Jsed for HgII (16.2 ± 7.3 ng m^{-2} day^{-1}) and MMHg (35.7 ± 17.4 ng m^{-2} day^{-1}) were higher during the dry season compared to the wet one (13.3 ± 9.1 ng m^{-2} day^{-1} and 5.4 ± 7.8 ng m^{-2} day^{-1} for HgII and MMHg, respectively). These MMHg diffusive fluxes are much lower than those measured in the carbonate-rich sediments of Lake Titicaca (36 to 224 ng m^{-2} day^{-1}) but in the range of those reported for the organic rich sediment of the same lake (13 to 60 ng m^{-2} day^{-1} [2]) and in coastal environments [3,4,6].

Strikingly, MMHg Jseds were higher than those of HgII during the dry season, whereas the opposite was found for the wet season. This greatly illustrates the major MMHg accumulation in the surface sediment PW and the upward MMHg efflux towards SW during the dry season. In addition, the positive correlation found between sulfide and MMHg Jsed (R = 0.70, $p < 0.01$, Figure S3) supports the notion that the upward shift of the redox gradient above the SWI during the dry season has likely favored the diffusion of DOC-bound MMHg or MMHg-SR complexes in the overlying water [14]. In contrast, the lower MMHg Jseds during the wet season likely result from the downward shift of the redox stratification in the epibenthic layers and/or the sediment. Indeed, higher MMHg Jsed during the night-time could be explained by a shallower redox front when the aquatic ecotope respiration dominates and favors upward Hg diffusion [2,44]. In contrast, the macrophytes and algae photosynthesis during the day may favor the water column oxygenation and the scavenging of MMHg in the Fe redox loop and onto submerged macrophytes' periphyton or benthic biofilms and algae [45,46], explaining negative Jsed during the day-time [4,47].

Hence, the stratification within the water column during the dry period likely favored upward MMHg effluxes from the sediment, which was also probably enhanced by the advection of surface PW by wind-induced shear stress in this shallow system as wind was on average 40% higher in November than in May (Figure S4).

4.2. Diel and Seasonal Changes in the Hg Partitioning between Dissolved and Particulate Phases in SW

Previous studies reported limited net Hg methylation potentials in the water column, whereas elevated ones were found in floating bio-organic substrates and surface sediment of Lake Uru Uru [18], supporting elevated MMHg production into fully anoxic environments or surface water microenvironments [12,48,49]. Although average MMHg concentrations were similar between the two seasons, the lower average percentages of filtered and unfiltered MMHg/THg with larger diel variations during the dry season compared to the wet one (i.e., 42 ± 4% and 35 ± 6%) can be explained by changes in water column biogeochemical conditions (i.e., dissolved oxygen, temperature and redox oscillations, and pelagic productivity or mobility), resulting in changes in the partitioning between the suspended solids and the solution. Alternatively, changes in the incident light radiation (see Section 4.3) between the two seasons may explain the seasonal differences.

During the wet season, average percentages of filtered (<0.22 µm) THg and MMHg in surface water were 58 ± 21% and 71 ± 23% of the UNF one, respectively, highlighting that both species were mainly present in the truly dissolved or colloidal phases (Figure 4c). Strikingly, sharp drops in the percentage of F/UNF THg and MMHg fractions (i.e., from ~80 to ~30%, Figure 4c) were found during the night, counterbalanced by the rise in both species onto particles synchronously with the drop in redox to negative values (Figure 2b). These percentages of F/UNF THg and MMHg were found to decrease with rising sulfides (R = −0.64 and −0.62, $p < 0.05$, for THg and MMHg, respectively) or DOC concentrations (R = −0.62 and −0.55, $p < 0.05$, for THg and MMHg, respectively) supporting the notion that DOC-bound MMHg or MMHg-SR complexes released from surface PW or produced in anoxic niches were sorbed onto suspended inorganic and organic particles. During the onset of reducing conditions, Mn concentrations (Figure S1) rose in the water column as a result of the dissolution of Mn oxides [50], whereas chalcophile redox-sensitive elements (i.e., Fe and V; Figure S1) decreased, likely due to their (co)precipitation as or onto authigenic sulfide minerals [51]. Sulfate concentrations (5.4 ± 5.1 g L^{-1}) were at least an order of magnitude higher than those of Fe, and thus the production of sulfides drives the Fe levels in the SW. Hence, drops in THg_F and $MMHg_F$ levels during reducing conditions found in the night-time period likely resulted from their adsorption or (co)precipitation onto/with colloidal and particulate FeS [2], and onto sulfurized OM [52]. They likely enriched the particulate pool through the flocculation of dissolved OM and the aggregation of colloids when SW became reducing [53,54].

During the dry season, Hg species in the reducing SW showed the opposite trend as for the wet season. THg and MMHg were mostly bound to particles (60 ± 13% and 50 ± 25% for particulate THg and MMHg, respectively), with higher partitioning onto particles during the day (Figure 4c). Such an opposite trend likely results from the more reducing, oxygen depleted, and sulfide-rich SWs, which favored both the MMHg production, upward efflux from the PW, and its preservation in the truly dissolved or colloidal form during the night in the dry season. In particular, the drop in oxygen was more accentuated during the night in the dry compared to the wet season, possibly resulting from a higher decomposition activity, triggered in turn by a higher productivity during the day. In addition, the higher content of algae and bacteria (both living or dead) in the SW during the dry season could have enhanced Hg and MMHg entrapment during the day as these organisms are known to accumulate MMHg in lake water columns [28,45,46,55]. Indeed, higher radiation levels and SW temperatures have increased biological activities, as illustrated by higher decreases in oxygen saturation (Figure 2a), which could have led to higher active uptake of MMHg by algae during the day [56]. Again, this agrees with reported abundant bio-organic aggregates and/or suspended particles enriched in OM in the water column during the dry season in Lake Uru Uru [18]. Thus, during the nights of the dry season, more MMHg is produced, whereas more uptake and potential demethylation occurs during the day. In addition, algae activity could have facilitated Hg reduction [57], explaining an increase in DGM during the day in the dry season (Figure 2a).

4.3. Solar Radiation Influence on MMHg Photodegradation and DGM Production in Surface Water

Because solar ultraviolet (UV-B) increases by around 10 ± 20% for every 1000 m increase in elevation [58,59], UV radiation intensity at Lake Uru Uru (3686 m a.s.l) is expected to be at least 40 to 80% higher than that measured at sea level [18,60]. The large diurnal and seasonal temperature amplitude gradients between the dry season (12.2 to 20.7 °C) and the wet one (10.0 to 13.3 °C, Figure 2a) broadly reflect these gradients in solar radiation.

For both seasons, filtered and unfiltered MMHg/THg decreased with increasing temperature, with the exception of the filtered phase during the wet season (Figure 5). This trend illustrates the strong influence of UV radiation on MMHg photodegradation [61] at both the diel and seasonal scales, with enhanced degradation of MMHg during the day and the dry season [18]. One cannot exclude higher activation of demethylating bacteria at higher temperatures [62].

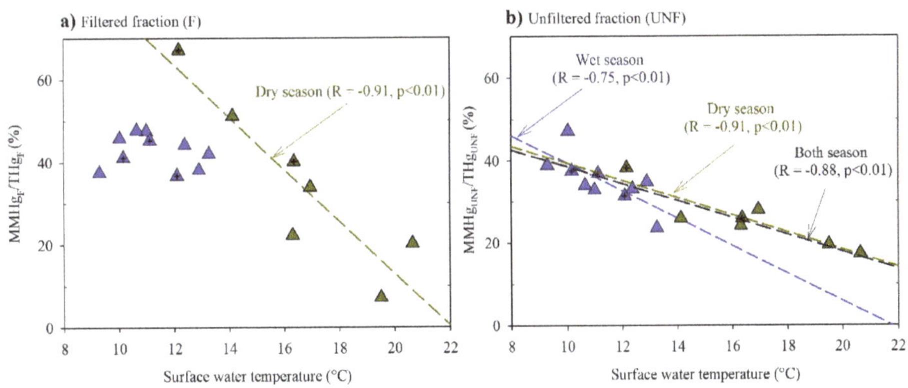

Figure 5. Plots of percentage MMHg/THg in (**a**) the filtered and (**b**) the unfiltered phases vs. surface water temperature. Crossed symbols indicate samples collected during the night-time period.

In addition, a positive correlation was found between DOC concentrations and percentages MMHg/THg (R = 0.66, $p < 0.01$) during the dry season, supporting the notion that rising DOC concentrations decreased the efficiency of MMHg photodegradation [63,64]. Although DOC concentration was lower during the wet season, the absence of a relationship between MMHg and SW temperature or DOC for the filtered fraction suggests that photodemethylation was not a primary sink for MMHg during this season. Indeed, lower insolation, higher water column, possible changes in the composition of photo-reactive DOC species, and/or different MMHg-DOC bindings [65] could have attenuated the radiation effect and reduced MMHg photodegradation efficiency. In particular, higher molecular weight dissolved OM, and a lower abundance of thiol functional groups (RS-) during the wet season compared to the dry one, could have attenuated MMHg photodegradation [66,67]. As mentioned previously, one cannot exclude the possibility of lower biological activity in the wet compared to the dry season, which would also have resulted in lower MMHg biodegradation.

As for MMHg, the overall low DGM (i.e., Hg(0)) concentrations in SW corroborate results from in situ incubations with Hg-enriched isotopes, which reported low biological and photochemical production of DGM in Lake Uru Uru [18]. Although DGM concentrations were low, the higher concentrations found during the dry season likely result from higher solar radiation, higher concentrations of suspended solids, as well as reducing bacteria/algae. Although solar radiations are the main fuel for DGM production during the day, rises in DGM found during the night support the notion of significant bacterial reductions in the dark [57].

5. Conclusions

Our results confirm previous observations made in Lake Uru Uru showing that diel changes in MMHg concentration overwhelm seasonal ones, with higher MMHg accumulation in surface water during the night-time and the dry season. We here bring new insights into the key role of the sediment compartment as a major source of MMHg for the SW, with a larger contribution in MMHg efflux towards the water column during the dry season when the lake is relatively shallow, more eutrophicated, and with the redoxcline likely moving above the SWI. Our results suggest that MMHg accumulation in the PWs fluctuates depending on diel redox oscillations around the SWI, driving both the bacterial sulfate reduction and bio-methylation. We here confirm the previous hypothesis from enriched isotope incubation that MMHg is preferentially produced in the PW and supports its release toward the SW associated with dissolved/colloidal OM or organic thiols. In the water column, both the pelagic light-dependent organisms and the flocculation/aggregation of dissolved OM with colloids drive the MMHg partitioning between dissolved and particulate phases during diel redox oscillations. Finally, although elevated MMHg diffusive fluxes demonstrate that MMHg loading from the porewater to the water column is large, both the photochemical and biological methylmercury degradation control the net MMHg accumulation in the water column.

Supplementary Materials: The following are available online at http://www.mdpi.com/2076-3417/10/21/7936/s1, Figure S1: Seasonal and diel cycles for vanadium (V), manganese (Mn), calcium (Ca); potassium (K), magnesium (Mg), sodium (Na), total sulfur (S), and silicon (Si) in surface water. Figure S2: Scatter plot with linear regression for porewater concentrations of (a) manganese (Mn) vs. hydrogen sulfide (H_2S) and dissolved oxygen (O2—measured in epibenthic water above the SWI with a multiparameter probe); (b) iron (Fe) and vanadium (V) vs. H_2S; (c) Filtered total mercury THg vs. H_2S and V. Linear regression was performed for all data (both seasons). Figure S3: Scatter plot for (upper panels) PW MMHg concentrations vs. hydrogen sulfide (H_2S) and vanadium (V), and (bottom panels) MMHg diffusive fluxes (Jsed MMHg) vs. (H_2S) and redox potential (measured in epibenthic water above the SWI with a multiparameter probe). Figure S4: Average precipitation and wind velocity at Oruro during the period 2010 to 2016 [68].

Author Contributions: Conceptualization, D.A. (David Amouroux), S.G., D.P., D.A. (Dario Achá), and S.B.; methodology, S.G., D.A. (David Amouroux), D.P., S.B., and D.A. (Dario Achá); validation, S.G., D.A. (Dario Achá), D.P., S.B., and D.A. (David Amouroux); methodology, S.G., D.A. (Dario Achá), D.P., S.B., and D.A. (David Amouroux); formal analysis, S.G., D.A. (Dario Achá), S.B., D.P., E.T., S.R.-L., C.H., P.F.-S., M.F., and S.B.; Field investigation, S.G., D.A. (David Amouroux), D.A. (Dario Achá), S.B., D.P., E.T., S.R.-L., C.H., P.F.-S., M.F., I.Q.-L., and S.B.; data curation, S.G.; writing—original draft preparation, S.G.; writing—review and editing, S.G., D.A. (Dario Achá), S.B., and D.A. (David Amouroux); supervision, S.G., D.A. (Dario Achá), S.B., and D.A. (David Amouroux); project administration, D.A. (David Amouroux) and S.G.; funding acquisition, S.G., D.A. (Dario Achá), S.B., and D.A. (David Amouroux). All authors have read and agreed to the published version of the manuscript.

Funding: This research was funded by Agence National pour la Recherche, as a contribution to the LA PACHÁMAMA project (ANR CESA program, No ANR-13-CESA-0015-01, PI: D. Amouroux: david.amouroux@univ-pau.fr), COMIBOL project (INSU CNRS/IRD EC2CO Program, PI: D. Point: david.point@ird.fr), EUTITICACA project (founded by the Impuestos Directos a los Hidrocarburos IDH administrated by the Universidad Mayor de San Andrés, PI: D. Achá: darioAchá@yahoo.ca) and TRACISOMER supported by a grant from Labex OSUG@2020 (PI: S. Guédron: stephane.guedron@ird.fr). S. Guédron and S. Bureau (ISTerre/IRD/UGA) are part of Labex OSUG@2020 (Investissements d'avenir—ANR10 LABX56).

Acknowledgments: We wish to thank J. Gardon, A. Terrazas, C. González, N. Clavijo, L. Salvatierra, R. Rios, J.C. Salinas, A. Castillo, M. Claure, and Don German Calizaya (Fishermen Association, MAchácamarca, Bolivia) for their help and assistance during the field campaigns.

Conflicts of Interest: The authors declare no conflict of interest.

References

1. Fitzgerald, W.F.; Lamborg, C.H.; Turekian, H.D.; Holland, K.K. 11.4 - Geochemistry of Mercury in the Environment. In *Treatise on Geochemistry*, 2nd ed.; Elsevier: Oxford, UK, 2014; pp. 91–129. [CrossRef]
2. Guédron, S.; Audry, S.; Achá, D.; Bouchet, S.; Point, D.; Condom, T.; Heredia, C.; Campillo, S.; Baya, P.A.; Groleau, A.; et al. Diagenetic production, accumulation and sediment-water exchanges of methylmercury in contrasted sediment facies of Lake Titicaca (Bolivia). *Sci. Total Environ.* **2020**, *723*, 138088. [CrossRef] [PubMed]
3. Guédron, S.; Huguet, L.; Vignati, D.A.L.; Liu, B.; Gimbert, F.; Ferrari, B.J.D.; Zonta, R.; Dominik, J. Tidal cycling of mercury and methylmercury between sediments and water column in the Venice Lagoon (Italy). *Mar. Chem.* **2012**, *130–131*, 1–11.
4. Bouchet, S.; Amouroux, D.; Rodriguez-Gonzalez, P.; Tessier, E.; Monperrus, M.; Thouzeau, G.; Clavier, J.; Amice, E.; Deborde, J.; Bujan, S. MMHg production and export from intertidal sediments to the water column of a tidal lagoon (Arcachon Bay, France). *Biogeochemistry* **2013**, *114*, 341–358.
5. Benoit, J.M.; Shull, D.H.; Harvey, R.M.; Beal, S.A. Effect of bioirrigation on sediment - water exchange of methylmercury in Boston Harbor, Massachusetts. *Environ. Sci. Technol.* **2009**, *43*, 3669–3674. [PubMed]
6. Hammerschmidt, C.R.; Fitzgerald, W.F. Sediment—Water exchange of methylmercury determined from shipboard benthic flux chambers. *Mar. Chem.* **2008**, *109*, 86–97. [CrossRef]
7. Hamelin, S.; Amyot, M.; Barkay, T.; Wang, Y.; Planas, D. Methanogens: Principal methylators of mercury in lake periphyton. *Environ. Sci. Technol.* **2014**, *45*, 7693–7700.
8. Achá, D.; Iniguez, V.; Roulet, M.; Guimarães, J.R.D.; Luna, R.; Alanoca, L.; Sanchez, S. Sulfate-Reducing Bacteria in Floating Macrophyte Rhizospheres from an Amazonian Floodplain Lake in Bolivia and Their Association with Hg Methylation. *Appl. Environ. Microbiol.* **2005**, *71*, 7531–7535. [CrossRef]
9. Garcia Bravo, A.; Bouchet, S.; Guédron, S.; Amouroux, D.; Dominik, J.; Zopfi, J. High methylmercury production under iron-reducing conditions in sediments impacted by sewage treatment plant discharges. *Water Res.* **2015**, *80*, 245–255.
10. Kerin, E.J.; Gilmour, C.C.; Roden, E.; Suzuki, M.T.; Coates, J.D.; Mason, R.P. Mercury methylation by dissimilatory iron-reducing bacteria. *Appl. Environ. Microbiol.* **2006**, *72*, 7919–7921. [CrossRef] [PubMed]
11. Point, D.; Monperrus, M.; Tessier, E.; Amouroux, D.; Chauvaud, L.; Thouzeau, G.; Jean, F.; Amice, E.; Grall, J.; Leynaert, A.; et al. Biological control of trace metal and organometal benthic fluxes in a eutrophic lagoon (Thau Lagoon, Mediterranean Sea, France). *Estuar. Coast. Shelf Sci.* **2007**, *72*, 457–471. [CrossRef]
12. Guédron, S.; Cossa, D.; Grimaldi, M.; Charlet, L. Methylmercury in tailings ponds of Amazonian gold mines (French Guiana): Field observations and an experimental flocculation method for *in situ* remediation. *Appl. Geochem.* **2011**, *26*, 222–229. [CrossRef]
13. Choe, K.Y.; Gill, G.A.; Lehman, R.D.; Han, S.; Heim, W.A.; Coale, K.H. Sediment—Water exchange of total mercury and monomethyl mercury in the San Francisco Bay - Delta. *Limnol. Oceanogr.* **2004**, *49*, 1512–1527. [CrossRef]
14. Skyllberg, U. Competition among thiols and inorganic sulfides and polysulfides for Hg and MeHg in wetland soils and sediments under suboxic conditions: Illumination of controversies and implications for MeHg net production. *J. Geophys. Res. Biogeosci. (2005–2012)* **2008**, *113*, 1–14. [CrossRef]
15. Bouchet, S.; Goni-Urriza, M.; Monperrus, M.; Guyoneaud, R.; Fernandez, P.; Heredia, C.; Tessier, E.; Gassie, C.; Point, D.; Guédron, S.; et al. Linking microbial activities and low molecular weight thiols to Hg methylation in biofilms and periphyton from high altitude tropical lakes (Bolivian altiplano). *Environ. Sci. Technol.* **2018**, *52*, 9758–9767. [CrossRef] [PubMed]

16. Guédron, S.; Devin, S.; Vignati, D.A.L. Total and methylmercury partitioning between colloids and true solution: From case studies in sediment overlying and porewaters to a generalized model. *Environ. Toxicol. Chem.* **2016**, *35*, 330–339. [CrossRef] [PubMed]
17. Helbling, E.W.; Villafañe, V.; Buma, A.; Andrade, M.; Zaratti, F. DNA damage and photosynthetic inhibition induced by solar ultraviolet radiation in tropical phytoplankton (Lake Titicaca, Bolivia). *Eur. J. Phycol.* **2001**, *36*, 157–166. [CrossRef]
18. Alanoca, L.; Amouroux, D.; Monperrus, M.; Tessier, E.; Goni, M.; Guyoneaud, R.; Achá, D.; Gassie, C.; Audry, S.; Garcia, M.; et al. Diurnal variability and biogeochemical reactivity of mercury species in an extreme high altitude lake ecosystem of the Bolivian Altiplano. *Environ. Sci. Pollut. Res.* **2016**, *23*, 6919–6933. [CrossRef]
19. Alanoca, L.; Guédron, S.; Amouroux, D.; Audry, S.; Monperrus, M.; Tessier, M.; Goix, S.; Achá, D.; Seyler, P.; Point, D. Synergistic effects of mining and urban effluents on the level and distribution of methylmercury in a shallow aquatic ecosystem of the Bolivian Altiplano. *Environ. Sci. Process. Impacts* **2016**, *18*, 1550–1560. [CrossRef]
20. Tapia, J.; Audry, S.; van Beek, P. Natural and anthropogenic controls on particulate metal(loid) deposition in Bolivian highland sediments, Lake Uru Uru (Bolivia). *Holocene* **2020**, *30*, 428–440. [CrossRef]
21. Sarret, G.; Guédron, S.; Achá, D.; Bureau, S.; Arnaud-Godet, F.; Tisserand, D.; Goni-Urriza, M.; Gassie, C.; Duwig, C.l.; Proux, O.; et al. Extreme Arsenic Bioaccumulation Factor Variability in Lake Titicaca, Bolivia. *Sci. Rep.* **2019**, *9*, 10626. [CrossRef]
22. Tapia, J.; Audry, S. Control of early diagenesis processes on trace metal (Cu, Zn, Cd, Pb and U) and metalloid (As, Sb) behaviors in mining-and smelting-impacted lacustrine environments of the Bolivian Altiplano. *Appl. Geochem.* **2013**, *31*, 60–78. [CrossRef]
23. Guédron, S.; Grimaldi, M.; Grimaldi, C.; Cossa, D.; Tisserand, D.; Charlet, L. Amazonian former gold mined soils as a source of methylmercury: Evidence from a small scale watershed in French Guiana. *Wat. Res.* **2011**, *45*, 2659–2669. [CrossRef]
24. Guédron, S.; Point, D.; Achá, D.; Bouchet, S.; Baya, P.A.; Molina, C.I.; Tessier, E.; Monperrus, M.; Flores, M.; Fernandez Saavedra, P.; et al. Mercury contamination level and speciation inventory in the hydrosystem of Lake Titicaca: Current status and future trends. *Environ. Pollut.* **2017**, *231*, 262–270. [CrossRef] [PubMed]
25. Bloom, N.S.; Fitzgerald, W.F. Determination of volatile mercury species at the picogram level by low-temperature gas chromatography with cold-vapor atomic fluorescence detection. *Anal. Chim. Acta* **1988**, *208*, 151–161. [CrossRef]
26. Cossa, D.; Averty, B.; Pirrone, N. The origin of methylmercury in open Mediterranean waters. *Limnol. Oceanogr.* **2009**, *54*, 837–844. [CrossRef]
27. Guédron, S.; Duwig, C.; Prado, B.L.; Point, D.; Flores, M.G.; Siebe, C. (Methyl) Mercury, Arsenic, and Lead Contamination of the World's Largest Wastewater Irrigation System: The Mezquital Valley (Hidalgo State-Mexico). *Water Air Soil Pollut.* **2014**, *225*, 1–19. [CrossRef]
28. Achá, D.; Guédron, S.; Amouroux, D.; Point, D.; Lazarro, X.; Fernandez, P.; Sarret, G. Algal bloom exacerbates hydrogen sulfide and methylmercury contamination in the emblematic high-altitude Lake Titicaca. *Geosciences* **2018**, *8*, 438. [CrossRef]
29. Monperrus, M.; Gonzalez, P.R.; Amouroux, D.; Alonso, J.I.G.; Donard, O.F.X. Evaluating the potential and limitations of double-spiking species-specific isotope dilution analysis for the accurate quantification of mercury species in different environmental matrices. *Anal. Bioanal. Chem.* **2008**, *390*, 655–666. [CrossRef]
30. Sharif, A.; Monperrus, M.; Tessier, E.; Bouchet, S.; Pinaly, H.; Rodriguez-Gonzalez, P.; Maron, P.; Amouroux, D. Fate of mercury species in the coastal plume of the Adour River estuary (Bay of Biscay, SW France). *Sci. Total Environ.* **2014**, *496*, 701–713. [CrossRef]
31. Small, J.M.; Hintelmann, H. Methylene blue derivatization then LC-MS analysis for measurement of trace levels of sulfide in aquatic samples. *Anal. Bioanal. Chem.* **2007**, *387*, 2881–2886. [CrossRef]
32. Small, J.M.; Hintelmann, H. Sulfide and mercury species profiles in two Ontario boreal shield lakes. *Chemosphere* **2014**, *111*, 96–102. [CrossRef] [PubMed]
33. Berner, R.A. *Early Diagenesis*; Princeton University Press: Princeton, NJ, USA, 1980; p. 240.
34. Boudreau, B.P. The diffusive tortuosity of fine-grained unlithified sediments. *Geochim. Cosmochim. Acta* **1996**, *60*, 3139–3142. [CrossRef]

35. Rothenberg, S.E.; Ambrose, R.F.; Jay, J.A. Mercury cycling in surface water, pore water and sediments of Mugu Lagoon, CA, USA. *Environ. Pollut.* **2008**, *154*, 32–45. [CrossRef]
36. Dill, C.; Kuiken, T.; Zhang, H.; Ensor, M. Diurnal variation of dissolved gaseous mercury (DGM) levels in a southern reservoir lake (Tennessee, USA) in relation to solar radiation. *Sci. Total Environ.* **2006**, *357*, 176–193. [CrossRef]
37. Amyot, M.; Mierle, G.; Lean, D.; McQueen, D.J. Effect of solar radiation on the formation of dissolved gaseus mercury in temperate lakes. *Geochim. Cosmochim. Acta* **1997**, *61*, 975–987. [CrossRef]
38. Soerensen, A.L.; Schartup, A.T.; Gustafsson, E.; Gustafsson, B.G.; Undeman, E.; Björn, E. Eutrophication Increases Phytoplankton Methylmercury Concentrations in a Coastal Sea -¸ A Baltic Sea Case Study. *Environ. Sci. Technol.* **2016**, *50*, 11787–11796. [CrossRef]
39. Lei, P.; Nunes, L.M.; Liu, Y.-R.; Zhong, H.; Pan, K. Mechanisms of algal biomass input enhanced microbial Hg methylation in lake sediments. *Environ. Int.* **2019**, *126*, 279–288. [CrossRef]
40. Knorr, K.H. DOC-dynamics in a small headwater catchment as driven by redox fluctuations and hydrological flow paths—Are DOC exports mediated by iron reduction/oxidation cycles? *Biogeosciences* **2013**, *10*, 891–904. [CrossRef]
41. Feyte, S.P.; Gobeil, C.; Tessier, A.; Cossa, D. Mercury dynamics in lake sediments. *Geochim. Cosmochim. Acta* **2012**, *82*, 92–112. [CrossRef]
42. Hellal, J.; Guédron, S.; Huguet, L.; Schaefer, J.; Laperche, V.; Joulian, C.; Lanceleur, L.; Burnol, A.; Ghestem, J.-P.; Garrido, F.; et al. Mercury mobilization and speciation linked to bacterial iron oxide and sulfate reduction: A column study to mimic reactive transfer in an anoxic aquifer. *J. Contam. Hydrol.* **2015**, *180*, 56–68. [CrossRef]
43. Benoit, J.M.; Gilmour, C.C.; Mason, R.P. Aspects of bioavailability of mercury for methylation in pure cultures of *Desulfobulbus propionicus* (1pr3). *Appl. Environ. Microbiol.* **2001**, *67*, 51–58. [CrossRef]
44. Gill, G.A.; Bloom, N.S.; Cappellino, S.; Driscoll, C.T.; Dobbs, C.; McShea, L.; Mason, R.; Rudd, J.W.M. Sediment-water fluxes of mercury in Lavaca Bay, Texas. *Environ. Sci. Technol.* **1999**, *33*, 663–669. [CrossRef]
45. Pickhardt, P.C.; Fisher, N.S. Accumulation of Inorganic and Methylmercury by Freshwater Phytoplankton in Two Contrasting Water Bodies. *Environ. Sci. Technol.* **2007**, *41*, 125–131. [CrossRef]
46. Quiroga-Flores, R.; Guédron, S.; Achá, D. High methylmercury uptake by green algae in Lake Titicaca: Potential implications for remediation. *Ecotoxicol. Environ. Saf.* **2020**, *207*, 111256. [CrossRef]
47. Point, D.; Bareille, G.; Pinaly, H.; Belin, C.; Donard, O.F.X. Multielemental speciation of trace elements in estuarine waters with automated on-site UV photolysis and resin chelation coupled to inductively coupled plasma mass spectrometry. *Talanta* **2007**, *72*, 1207–1216. [CrossRef] [PubMed]
48. Achá, D.; Pabon, C.A.; Hintelmann, H. Mercury methylation and hydrogen sulfide production among unexpected strains isolated from periphyton of two macrophytes of the Amazon. *FEMS Microbiol. Ecol.* **2012**, *80*, 637–645. [CrossRef]
49. Gascon Diez, E.; Loizeau, J.-L.; Cosio, C.; Bouchet, S.; Adatte, T.; Amouroux, D.; Bravo, A.G. Role of settling particles on mercury methylation in the oxic water column of freshwater systems. *Environ. Sci. Technol.* **2016**, *50*, 11672–11679. [CrossRef]
50. Audry, S.; Blanc, G.; Schäfer, J.; Chaillou, G.N.L.; Robert, S.B. Early diagenesis of trace metals (Cd, Cu, Co, Ni, U, Mo, and V) in the freshwater reaches of a macrotidal estuary. *Geochim. Cosmochim. Acta* **2006**, *70*, 2264–2282. [CrossRef]
51. Wanty, R.B.; Goldhaber, M.B. Thermodynamics and kinetics of reactions involving vanadium in natural systems: Accumulation of vanadium in sedimentary rocks. *Geochim. Cosmochim. Acta* **1992**, *56*, 1471–1483. [CrossRef]
52. Graham, A.M.; Cameron-Burr, K.T.; Hajic, H.A.; Lee, C.; Msekela, D.; Gilmour, C.C. Sulfurization of Dissolved Organic Matter Increases Hg-Sulfide-Dissolved Organic Matter Bioavailability to a Hg-Methylating Bacterium. *Environ. Sci. Technol.* **2017**, *51*, 9080–9088. [CrossRef]
53. Von Wachenfeldt, E.; Tranvik, L.J. Sedimentation in Boreal Lakes: The Role of Flocculation of Allochthonous Dissolved Organic Matter in the Water Column. *Ecosystems* **2008**, *11*, 803–814. [CrossRef]
54. Yan, J.; Lazouskaya, V.; Jin, Y. Soil colloid release affected by dissolved organic matter and redox conditions. *Vadose Zone J.* **2016**, *15*, 1–10. [CrossRef]
55. Lanza, W.G.; Achá, D.; Point, D.; Masbou, J.; Alanoca, L.; Amouroux, D.; Lazzaro, X. Association of a Specific Algal Group with Methylmercury Accumulation in Periphyton of a Tropical High-Altitude Andean Lake. *Arch. Environ. Contam. Toxicol.* **2017**, *72*, 1–10. [CrossRef]

56. Moye, H.A.; Miles, C.J.; Phlips, E.J.; Sargent, B.; Merritt, K.K. Kinetics and Uptake Mechanisms for Monomethylmercury between Freshwater Algae and Water. *Environ. Sci. Technol.* **2002**, *36*, 3550–3555. [CrossRef]
57. Deng, L.; Fu, D.; Deng, N. Photo-induced transformations of mercury(II) species in the presence of algae, Chlorella vulgaris. *J. Hazard. Mater.* **2009**, *164*, 798–805. [CrossRef]
58. Andrade, M.; Forno, R.; Palenque, E.; Zaratti, F. Estudio preliminar del efecto de altura sobre la radiación solar ultravioleta B. *Rev. Bol. Fis.* **1998**, *4*, 14.
59. Blumthaler, M.; Ambach, W.; Rehwald, W. Solar UV-A and UV-B radiation fluxes at two alpine stations at different altitudes. *Theor. Appl. Climatol.* **1992**, *46*, 39–44. [CrossRef]
60. Villafae, V.E.; Andrade, M.; Lairana, V.; Zaratti, F.; Helbling, E.W. Inhibition of phytoplankton photosynthesis by solar ultraviolet radiation: Studies in Lake Titicaca, Bolivia. *Freshw. Biol.* **1999**, *42*, 215–224. [CrossRef]
61. Klapstein, S.J.; O'Driscoll, N.J. Methylmercury biogeochemistry in freshwater ecosystems: A review focusing on DOM and photodemethylation. *Bull. Environ. Contam. Toxicol.* **2018**, *100*, 14–25. [CrossRef]
62. Oremland, R.S.; Culbertson, C.W.; Winfrey, M.R. Methylmercury decomposition in sediments and bacterial cultures: Involvement of methanogens and sulfate reducers in oxidative demethylation. *Appl. Environ. Microbiol.* **1991**, *57*, 130–137. [CrossRef] [PubMed]
63. Klapstein, S.J.; Ziegler, S.E.; O'Driscoll, N.J. Methylmercury photodemethylation is inhibited in lakes with high dissolved organic matter. *Environ. Pollut.* **2018**, *232*, 392–401. [CrossRef] [PubMed]
64. Klapstein, S.J.; Ziegler, S.E.; Risk, D.A.; O'Driscoll, N.J. Quantifying the effects of photoreactive dissolved organic matter on methylmercury photodemethylation rates in freshwaters. *Environ. Toxicol. Chem.* **2017**, *36*, 1493–1502. [CrossRef]
65. Jeremiason, J.D.; Portner, J.C.; Aiken, G.R.; Hiranaka, A.J.; Dvorak, M.T.; Tran, K.T.; Latch, D.E. Photoreduction of Hg (II) and photodemethylation of methylmercury: The key role of thiol sites on dissolved organic matter. *Environ. Sci. Process. Impacts* **2015**, *17*, 1892–1903. [CrossRef] [PubMed]
66. Ni, B.; Kramer, J.R.; Bell, R.A.; Werstiuk, N.H. Protonolysis of the Hgâ^′ C Bond of chloromethylmercury and dimethylmercury. A DFT and QTAIM Study. *J. Phys. Chem. A* **2006**, *110*, 9451–9458. [CrossRef] [PubMed]
67. Zhang, D.; Yin, Y.; Li, Y.; Cai, Y.; Liu, J. Critical role of natural organic matter in photodegradation of methylmercury in water: Molecular weight and interactive effects with other environmental factors. *Sci. Total Environ.* **2017**, *578*, 535–541. [CrossRef] [PubMed]
68. SENAMHI, B. Servicio Nacional de Meteorología de Hidrología. *La PazBoliv.* Available online: http://www.senamhi.gob.bo (accessed on 25 August 2020).

Publisher's Note: MDPI stays neutral with regard to jurisdictional claims in published maps and institutional affiliations.

applied sciences

MDPI

Article

Influence of Ecological Restoration on Mercury Mobility and Microbial Activities on Former Guyanese Mining Sites

Ewan Couic [1,*], Vanessa Alphonse [2], Alexandre Livet [2], Stéphanie Giusti-Miller [3] and Noureddine Bousserrhine [2,*]

1 Department of Geosciences, Observatory of Sciences of the Universe, UMR 6118 Géosciences, 35000 Rennes, France

2 Water, Environment and Urban Systems Laboratory (Leesu), Department of Biogeochemistry, University of Paris-Créteil, 94010 Créteil, France; vanessa.alphonse@u-pec.fr (V.A.); livet@u-pec.fr (A.L.)

3 Institute of Ecology and Environmental Sciences of Paris (IEES), University of Paris-Créteil, 94010 Créteil, France; giusti@u-pec.fr

* Correspondence: ewan.couic@univ-rennes1.fr (E.C.); bousserrhine@u-pec.fr (N.B.); Tel.: +33-6236-16195 (E.C.); +33-6498-58970 (N.B.)

Abstract: As rehabilitation efforts in Guyana are recent, there is little information on the effect of different ecological rehabilitation protocols for Guyana's mining sites on biogeochemical cycles and mercury mobility. This study was conducted to assess the impact of different ecological restoration protocols on soil quality with the use of soil microbial indicators and by estimating the mercury mobility. We sampled soil from six rehabilitated mining sites in French Guyana with different ecological restoration procedures. We carried out measurements of enzymatic activities and an analysis of mercury environmental speciation to assess its potential toxicity according to a mobility gradient. The results obtained in this study show that the rehabilitation of mining sites has been carried out in a heterogeneous manner and soil quality is very variable, even in nearby sites. Sites that have been rehabilitated with fabaceous species have positive soil quality indicators. In addition, the results highlight a change in mercury mobility that is 82.1% correlated after co-inertia analysis with soil texture properties, which also confirms a direct effect of rehabilitation on mercury mobility. The non-restored sites had a much higher potential of mercury mobility and toxicity than the sites where ecological restoration was successful. These results highlight the positive effect of controlled rehabilitation and ecological restoration on microbiological activities and the potential toxicity of mercury.

Keywords: ecological restoration; mercury mobility; microbial activities; biogeochemistry; gold mining activities; French Guiana

Citation: Couic, E.; Alphonse, V.; Livet, A.; Giusti-Miller, S.; Bousserrhine, N. Influence of Ecological Restoration on Mercury Mobility and Microbial Activities on Former Guyanese Mining Sites. *Appl. Sci.* **2021**, *11*, 2231. https://doi.org/10.3390/app11052231

Academic Editor: Stéphane Guédron

Received: 27 January 2021
Accepted: 26 February 2021
Published: 3 March 2021

1. Introduction

Deforestation is currently one of the sectors that emit the most greenhouse gases [1,2] and although forest areas are increasing globally, tropical forests are particularly affected by this crisis [3]. Among the causes of deforestation in the Amazon, agriculture, silviculture, cattle ranching, selective logging, coca farming, and artisanal scale gold mining (ASGM) are responsible for a large percentage of forest loss. In South America and in French Guiana in particular, the stakes of gold mining are serious for the development of the region, and are worrying for the preservation of the ecosystem. In Guyana, gold mining impacts land use and the functioning of the forest system, including the soil [4,5]. To exploit gold in alluvial terraces, it is necessary to remove all the vegetation and then to use powerful water jets (sluice) to eliminate the superficial layers of soil until reaching the gold-bearing layer [6]. In addition to the direct damage caused by deforestation, these practices lead to heavy soil losses through erosion, which will lead to increased turbidity in aquatic systems and the remobilization of toxic metallic trace elements such as mercury [7]. Due to amalgamating chemical properties for gold mining, mercury was widely and legally used until 2006 and the problem of mercury in Guiana has already been described in many scientific studies [8,9]. It was shown that mercury of anthropogenic origin in Guyana was the most reactive, highly disorganized and often had suboxic physico-chemical conditions of the mining sites, which are favored its mobility and methylation [10,11].

The biogeochemistry of mercury is complex and depends on many physico-chemical parameters (i.e., temperature, redox potential, pH, metal oxides, organic matter, and mercury oxidation level) [12], but also biological parameters. Depending on these conditions, the forms of mercury in the soil will be affected and this will influence its mobility, bioavailability and toxicity to biological systems [13]. Free forms of mercury take different forms in soils such as alkyl mercury species such as methylmercury ($MeHg^{+}$) or ethylmercury(II) ($EtHg^{+}$) or inorganic soluble mercury species. These forms generally represent the most toxic component of mercury due to their high mobility. Metal mercury complexes, or certain amalgams, are considered to be less toxic because they are more difficult to mobilize without specific reactions, such as chemical or microbial reduction. The least toxic forms of mercury, such as mercuric sulphide or cinnabar (HgS) [14,15], are generally stable in soils, but may be weathered by simple dissolution in acidic environments [15] by significant root exudates such as those from fabaceae [16] or by microbial production of low molecular weight organic acid (LMMOAs) [17]. Determining mercury speciation at anthropized sites such as former mining sites is therefore very important for assessing the positive or negative health impact of ecological rehabilitation [15].

In order to limit the adverse effects associated with mercury mobility, ecological rehabilitation methods have been proposed with varying degrees of success in the Guyana basin. In the mining code, there is a difference between the principle of rehabilitation and restoration or revegetation [18]. Rehabilitation in the mining sense consists solely of reconstituting the original soil from the materials that have been extracted and moved, terracing the mixture, and if possible, covering it with so-called fertile horizons. Restoration will consist of applying a defined site revegetation protocol, if possible using local species in order to rebuild the original ecosystem [19,20]. While rehabilitation is mandatory, restoration is not explicitly requested in the mining code and too often remains at the discretion of mining companies. Among the methods used to evaluate the quality of ecological restoration protocols, enzymatic activity measurements and quantification of soil microbial biomass are among the most effective and widely used methods, particularly at mining sites [21,22].

While data on mercury speciation in natural soils and mining sites are fairly well documented [9,23], changes in carrier phases can alter mercury mobility and these mechanisms during ecological restoration processes are still not well known in Guyana and are essential to define a chemical quality status of the soil. In a context where the physico-chemical conditions (iron oxides, low pH and suboxic conditions) of rehabilitated sites could promote mercury mobility [11], environmental speciation measures could therefore be an interesting tool for assessing the health quality of these plots. One of the additional effects of good ecological rehabilitation is also to limit anoxic soil conditions through better regulation of the water cycle. In the context of tropical soils that are potentially rich in mercury, ecological restoration could therefore limit the conditions conducive to mercury mobility. While the effect of ecological restoration has already been shown to be positive for soil biological quality [22,24], it would also be relevant to assess the impact of different vegetation cover on mercury speciation by the use of environmental extraction procedures [25].

The main assumptions led to his work on the quality of rehabilitation, which was linked to an increase in plant cover and soil stability and can limit mercury mobility and promote soil microbial activities. We also hypothesize that the toxic fractions of mercury in the soil have a negative impact on microbial communities. In this context, the main physico-chemical properties of the soil, telluric enzymatic activities indicative of the functioning of biogeochemical cycles, as well as the environmental speciation of mercury were determined in different soil samples from former mining sites rehabilitated in French Guiana, restored or not to see the effect of a vegetation gradient. The objectives of this study were: (1) assess the biological quality of mine sites in relation to the quality of rehabilitation and the level of ecological restoration; and (2) describe the influence of heterogeneous soil rehabilitation and different ecological restoration modalities on the fate and mobility of mercury in soils. To answer these questions, six sites undergoing ecological restoration have been selected in French Guiana, in the Yaoni and Belizon regions. These sites were chosen because of their similar gold mining liabilities, as well as for their ecological restoration gradients, both in terms of quality and age.

It is difficult to anticipate the different recovery trajectories of biogeochemical cycles on a wide range of rehabilitated mining sites; however, these results could contribute to improving the assessment of the sanitary quality of rehabilitated sites in French Guiana.

2. Materials and Methods

2.1. Site Description and Soil Sampling

The work was undertaken in French Guiana, South America. The sampled soils (Figure 1) for this study were rehabilitated after gold mining extraction. After closing the alluvional mine pit, rehabilitation consisted of reconstituting the soil excavated during mine exploitation.

Figure 1. Illustration of some examples of different stages of ecological restoration on formers Guyanese gold mining sites and geographical location of the Yaoni and Belizon mines in French Guiana. Non-restored site (**top left**), ecological restoration with herbaceous species after 3 years ((**top right**), Bel1), ecological restoration with fabaceous species after 3 years ((**bottom left**), Bel3), ecological restoration with fabaceous species after 18 years ((**bottom right**), Yao3). The pictures were taken during a second sampling campaign in April 2016 on sites undergoing ecological restoration.

The first three sites were in the Belizon forest (N 04°22.200′/W 052°19.328′). The mining operation lasted five years and ended in January 2013. All sites in the Belizon have had the same mining history. In February 2013, an ecological restoration programme was initiated with macrocuts of *Clitoria racemosa*, a species of local fabacea. The recovery of *C. racemosa* macrocuts has not been effective on all sites in the Belizon forest. On the Bel1 site, vegetation covered between 5% and 10% of the soil surface with a majority of cyperaceae species (*Cyperus* sp. and *Carex* sp.). On the Bel2 site, vegetation covered more than 70% of the site and almost the entire site is dominated by *Lycopodiella* sp. with the presence of some cyperaceae species. On the Bel3 site, macrocuts of *C. racemosa* were recovered and the trees are nearly 5 m high.

The following three sites were taken from the former Yaoni mine site (N 04°30.930′/W 052°21.275′). Mining activity at the Yaoni sites lasted several years and ended in 1997. These three sites also have the advantage of having the same mining history. The first two Yaoni sites (Yao1 and Yao2) have not been restored by specific protocols and the low vegetation cover, mainly *Lycopodiella* sp., *Cyperus* sp. and *Carex* sp., is the result of spontaneous vegetation development. The Yao3 site has been completely restored by associations of fabaceous species, notably *Acacia mangium* and *Clitoria racemosa*.

At each site, soil samples were collected from depths between 0–10 cm with an auger in April 2016. The sampling for the five sites consisted of sampling five plots ($n = 5$) per site, with three sub-samples per plot that were pooled at the laboratory. The sampling of five replicas per site was mainly conditioned by the topography of the terrain, the accessibility of the plots and the feasibility of sampling. Soil samples were immediately sealed in sterile hermetic polyethylene bags for transportation to the Cayenne IRD laboratory. Samples were then dried at ambient temperature (25 °C) until they were air dried (i.e., approximately three weeks). These 90 collected samples (6 sites × 5 plots × 3 samples per plot) were then sieved at 2 mm, homogenized, and hermetically sealed at 4 °C until used.

2.2. Soil Samples Characterization

Granulometry was determined in the fraction less than 2 mm, and five classes of particles have been distinguished according to the NF X 31-107 standard: clay (<2 µm), fine silts (2 to 20 µm), coarse silt (20 to 50 µm), fine sands (0.050 to 0.200 mm) and coarse sands (0.200 to 2 mm).

For the total Fe and Al analyses by inductively coupled plasma optical emission spectroscopy (ICP-OES, Spectroblue, Elancourt-France), the samples were first ground to 63 µm then an acid digestion was performed in cleaned Teflon digestion tubes (SCP Science, Villebon sur Yvette-France) using a hot block (Digiprep MS, SCP Science) with two digestion HNO_3 (VWR ref. 83872.330)-HCL (VWR ref. 83871.290) in a 1:3 ratio and HF (Sigma-Aldrich, Saint Louis, MO, USA, ref. 339261-100 ML).

Soil pH-H_2O and pH-KCl were measured in soil suspensions according to ISO 10390. Soil total carbon (Ctot) and total nitrogen (Ntot) were determined by the Dumas method (NF ISO 13878) by chromatography with a thermal conductivity detector (NA 1500 série 2 CARLO-ERBA). Total phosphorus (Ptot) was determined after the acid digestion of soil samples, by Murphy and Riley method [26]. Water extractions were performed for total organic carbon (TOC), and concentrations were measured with a Shimadzu TOC-500 apparatus (Shimadzu, Kyoto, Japan).

2.3. Soil Microbial Biomass and Enzyme Assays

The soil microbial biomass was estimated according to Anderson and Domsch's work [27]. Using the SIR method, soil microbial biomass C (SIR-biomass) was calculated using the following equation:

$$\text{SIR-biomass C } (\mu\text{gC g}^{-1}\text{ soil}) = \text{SIR } (\mu\text{L CO}_2\text{ g}^{-1}\text{ soil h}^{-1}) \times 40.04 + 0.37 \quad (1)$$

The activity of acid (AcdP) and alkaline (AlkP) phosphatases, B-glucosidase (Glu) and arylsulfatase (Aryl) were assayed based on the amount of p-nitrophenol (pNP) released after cleavage of enzyme-specific substrates at the average pH of natural soil. These enzymes were measured, as described by Badiane et al. and Tabatabay and Bremner [28–30]. Dehydrogenase activity (DHA) was determined according to Klein et al. [31] using 2.3.5_tripenyltetrazolium chloride (TTC) as substrate. Urease activity (Ur) was determined according to Kandeler and Gerber [32] using urea as substrate. The total enzymatic activity (FDA) of soil microorganisms was determined by a spectrophotometric method using fluoresceine diacetate as substrate following the method developed by Green et al. [33].

2.4. Mercury Analysis

2.4.1. Total Mercury Content

Mercury concentrations (HgT) in soil samples were determined directly by thermal decomposition atomic absorption spectrometry after gold amalgamation using an automatic mercury analyzer (AMA 254) in Ultra High Sensitivity configuration (ref 002-001-001) with a detection limit of 0.001 ng and quantification limit of 0.003 ng.

2.4.2. Environmental Speciation of Mercury

- Extraction of "Highly mobile" mercury species

This extraction included three fractions (water soluble, exchangeable and bioavailable mercury fraction) of the most mobile and bioavailable forms of mercury. The sum of the three fractions (F1 + F2 + F3) in the next three paragraphs will be called E1.

Soluble mercury (F1) was estimated according to Rasmussen et al. [34] with ultrapure water. The exchangeable Hg fractions (F2) in soil samples were measured according to Silveira et al. [35] with 0.1 M $CaCl_2$ (Sigma-Aldrich, ref. 223506-500 G). The bioavailable mercury contents (F3) in soil samples were measured after a 1:10 (w/v) soil/0.01 M EDTA (Carl Roth, Karlsruhe-Germany, ref. 8040.3) + 1 M ammonium acetate (Merck, Darmstadt-Germany, ref. 1.01115.1000) following the standard AFNOR NF X31-120. Then, solutions were centrifuged, supernatants were filtered at 0.45 mm with Teflon filters (Minisart SRP 25, Sartorius, Dourdan-France), and then 200 µL of extracted supernatants were analyzed for exchangeable mercury content with AMA-254.

- Extraction of toxic, semi-mobile and non-mobile mercury

To assess the environmental speciation of mercury in soil, we followed Han et al. [14] procedure in order to determine the mobile and toxic, semi-mobile and non-mobile fractions of mercury in the restored soils and all fractions were analyzed by the AMA-254.

Briefly, the extraction of mobile and toxic mercury, called E2, involves the use of a solution of 2% HCl + 10% ethanol to extract the mobile mercury species from the soil. The target mercury species include toxic alkyl mercury species, such as $MeHg^+$ and $EtHg^+$ species, as well as inorganic mercury species that have great mobility, such as soluble Hg^{2+}. As this extraction targeted all forms of mobile mercury, including the three fractions detailed above, extraction E2 corresponded to the difference between the total mobile mercury extracted by this protocol and the sum of the fractions of extraction E1.

To extract the "semi-mobile" mercury, called E3, species including mainly Hg and mercury-metal amalgam, the sample remaining after the ethanol extraction was first sonicated with 5 mL of ultrapure water at 60 °C for 5 min. The sample was centrifuged, and the supernatant was discarded. Then, extraction was carried out with 5 mL 1:2 (v/v) HNO_3: Ultrapure water. The rest of the matric material could be directly analyzed by AMA-254 for "non-mobile" mercury species, called E4, including mainly Hg_2Cl_2 and HgS mercury species.

2.5. Statistical Analysis

The normality of the data distributions and the equal variance between treatments were tested using the Shapiro test and Bartlett's test, respectively. To study the effect of the restoration processes, we conducted different one-way ANOVA. Tukey's HSD multiple comparison method was used to test the difference between the different restoration processes. Pearson correlation was performed between mercury speciation and the dataset of enzyme activities and soil properties.

Co-inertia analyses, a two-table ordination method, were used to analyze the impact of soil properties (physics and chemistry) on enzymes activities and mercury speciation. This involves a simultaneous projection, at the same scale, of the PCA (data not showed) conducted on soil properties and the PCA conducted on enzyme datasets and a mercury speciation dataset. Permutation tests were conducted to assess the statistical significance of the co-variation between the different datasets. The R software was used for all statistical analyses (R version 3.3.2 (31 October 2016)).

2.6. Quality Assurance and Control

To avoid contamination, all materials used in this work were acid-washed twice with HNO_3 (5%) and then rinsed several times with Milli-Q water before used.

Concerning mercury assays in samples, the relative error was routinely ±5% and was always under ±10%. The detection limit (defined as three times the standard deviation (SD) of the blank) was 0.03 µg kg^{-1}. Concentrations obtained for repeated analyses of certified reference materials never exceeded the published range of concentrations (i.e., 0.128 ± 0.017 µg g^{-1} and 0.091 ± 0.009 µg g^{-1} for BCR-277R and MESS-3, respectively).

3. Results

3.1. Soil Physical and Chemical Properties of the 6 Sites

Granulometry measurements (Table 1) showed heterogeneous textures between the rehabilitated sites. The Bel1, Bel3, Yao2 and Yao3 sites were predominantly sandy with significantly different fine sand/coarse sand ratios between the sites. The Bel2 site was mostly silty while the Yao1 site was mainly clayey. The soils rehabilitated after mining with heterogeneous structures and textures, can be considered as anthroposols.

Table 1. TOC: Total Organic Carbon, C_{tot}, N_{tot}, and P_{tot}: Total C, N and P, Total Fe, Al: Total Iron, Aluminum content, (n = 5, mean ± SD). For each parameter, values followed by different letters differ significantly with $p < 0.05$ with Tukey HSD test.

Sites Characteristics	Bel1	Bel2	Bel3	Yao1	Yao2	Yao3
TOC (g·kg^{-1})	2.47 ± 0.22 **a**	5.84 ± 0.34 **b**	4.99 ± 0.37 **b**	3.88 ± 0.03 **ab**	3.51 ± 0.23 **ab**	15.58 ± 1.4 **c**
C total (g·kg^{-1})	3.52 ± 0.36 **a**	9.54 ± 0.73 **b**	7.01 ± 0.83 **b**	6.01 ± 0.32 **b**	4.29 ± 0.19 **a**	22.88 ± 1.51 **c**
N total (g·kg^{-1})	0.14 ± 0.02 **a**	0.45 ± 0.01 **b**	0.40 ± 0.04 **b**	0.40 ± 0.02 **b**	0.28 ± 0.03 **ab**	1.81 ± 0.04 **c**
C/N	25.14	21.20	17.52	15.02	15.32	12.64
P total (g·kg^{-1})	0.13 ± 0.01 **a**	0.27 ± 0.07 **b**	0.17 ± 0.02 **ab**	0.21 ± 0.01 **ab**	0.18 ±0.01 **ab**	0.75 ± 0.03 **c**
pH	5.3 ± 0.1 **a**	4.66 ± 0.08 **b**	5.00 ± 0.03 **ab**	5.26 ± 0.2 **a**	5.02 ± 0.03 **ab**	4.66 ± 0.2 **b**
Clay%	7.05 ± 2.7 **a**	22.24 ± 2.4 **b**	21.26 ± 1.5 **b**	46.44 ± 1.1 **c**	24.26 ± 0.5 **b**	19.78 ± 1.4 **b**
Fin silt%	18.62 ± 2.1 **b**	44.47 ± 2.6 **d**	17.68 ± 3.5 **b**	33.92 ± 1.2 **c**	10.10 ± 0.3 **a**	27.7 ± 0.8 **c**
Coarse silt%	2.39 ± 0.98 **a**	21.39 ± 0.92 **d**	7.10 ± 1.1 **b**	7.57 ± 0.45 **b**	10.40 ± 0.7 **c**	7.8 ± 0.5 **b**
Fin sand%	63.68 ± 1.4 **e**	2.67 ± 1.1 **a**	4.86 ± 0.6 **a**	8.59 ± 0.49 **b**	35.26 ± 0.7 **d**	14.17 ± 1.35 **c**
Coarse sand%	7.41 ± 0.33 **b**	9.14 ± 0.8 **b**	49 ± 5 **d**	3.30 ± 0.48 **a**	19.30 ± 0.5 **c**	22.22 ± 1.17 **c**
Fe (g·kg^{-1})	66.85 ± 6.2 **b**	44.4 ± 2.8 **b**	113.3 ± 15.2 **a**	65.04 ± 10 **ab**	75.7 ± 12.9 **b**	71.4 ± 4.9 **b**
Al (g·kg^{-1})	16.54 ± 2.12 **a**	41.4 ± 2.2 **a**	25.6 ± 2.15 **b**	53.6 ± 15.8 **b**	76.4 ± 4.8 **b**	43.3 ± 10.2 **b**

Among the six rehabilitated sites, the two sites restored Bel3 and Yao3 had significantly higher levels of total CNP and TOC content than the non-restored sites. The total carbon content, nitrogen and phosphorus varied, respectively, between 3.52 and 22.88 g·kg^{-1}; 0.14 and 1.81 g·kg^{-1} and 0.13 and 0.75 g·kg^{-1} between non-restored and restored sites. The C/N ratio of the Bel1 and Bel2 sites could indicate significant nitrogen deficiencies limiting ecological restoration processes. On the contrary, the old Yaoni sites have lower C/N ratios, indicative of a better cycle of organic matter degradation.

The six sites were rich in iron with minor differences, which seems consistent with Guyanese soils [36–38]. Total iron content in soil varied between 44.4 and 85.04 g·kg^{-1} for Bel3 and Yao2, respectively. The aluminum contents were homogeneous for Bel3, Yao1, Yao2 and Yao3 sites and higher than Bel1 and Bel2 sites.

3.2. Soil Microbial Biomass and Enzyme Activities

The two sites restored with fabaceous species, Bel3 and Yao3 sites, had significantly higher soil microbial biomass levels than the non-restored sites (Table 2) with a maximum biomass of 1102 mg kg^{-1}.

Regardless of the soil enzyme assay (Table 2) and the rehabilitated sites, soil enzyme activities showed strong significant differences between restored and non-restored sites. The total enzymatic activities (FDA) varied between 0.49 and 3.22 µg·g^{-1}·h^{-1} of fluorescein for Yao2 and Yao3, respectively. For dehydrogenase activity, the maximum activity was for Yao3 site, while the minimum was for the Bel2 site with a value of 5.80 mg kg^{-1} and 0.34 µg TPF g^{-1} soil h^{-1}, respectively. For the β-glucosidase activity, the Yao3 site had significantly the highest activity with 18.6 µg pNP g^{-1} soil h^{-1}, while the other sites had a relatively homogeneous activity. For urease activity, the highest value was measured for Yao3, then for Bel3 with 4.74 and 1.52 µg $NH_4{}^+$-N g^{-1} soil h^{-1}, respectively. Acid phosphatase activities were higher than for the other enzymes and significantly higher for the two restored sites and Yao2 than for the other non-restored sites. For alkaline phosphatase activity, the highest value was measured for Yao2, then for Bel2 with 12.07 and 4.17 µg pNP g^{-1} soil h^{-1}, respectively. For Arylsulfatase activity, the Yao3 and Bel3 sites reached values of 56.0 and 8.42 µg pNP g^{-1} soil h^{-1}, respectively, while the other sites had very little activity.

The results, in particular the microbial biomass, illustrated the positive impact of ecological restoration on the return of microbial communities in rehabilitated soils, in direct correlation with a return of enzymatic activities responsible for the turnover of nutrients in the soil and the functioning of ecosystems. The establishment of fabaceae species seemed to be a very interesting choice as a pioneer species to be used on rehabilitated sites.

Table 2. Soil Microbial Biomass and soil enzymes activities involved in soil biogeochemical cycles in the six rehabilitated sites (n = 5, mean ± SD). MBC: Soil microbial biomass carbon in mg·kg^{-1}, DH: Dehydrogenase activity in µg TPF (red-colored formazan) g^{-1} soil h^{-1}, B-Glu, AcdP, AlkP and Aryl: β-glucosidase, Acid phosphatase, Alkaline phosphatase and Arylsulfatase activities in µg pNP g^{-1} soil h^{-1}, Urease: Urease activity in µg NH_4^+-N g^{-1} soil h^{-1}, FDA: Global microbial enzymes activities in µg fluorescein g^{-1} soil h^{-1}. For each parameter, values followed by different letters differ significantly with $p < 0.05$ with Tukey HSD test.

Microbial Activities	Bel1	Bel2	Bel3	Yao1	Yao2	Yao3
MBC	312.6 ± 10.0 **c**	385.9 ± 21.5 **c**	844.4 ± 14.1 **b**	223.1 ± 10.8 **d**	248.1 ± 13.6 **d**	1102 ± 41 **a**
DH	0.43 ± 0.08 **d**	0.34 ± 0.02 **d**	2.10 ± 0.19 **b**	0.46 ± 0.07 **d**	1.03 ± 0.12 **c**	5.80 ± 0.44 **a**
B-Glu	1.60 ± 0.25 **a**	1.46 ± 0.78 **a**	5.54 ± 0.38 **a**	1.38 ± 0.10 **a**	3.48 ± 0.97 **a**	18.6 ± 3.4 **b**
Urease	0.46 ± 0.04 **c**	0.70 ± 0.02 **bc**	1.52 ± 0.06 **b**	0.14 ± 0.01 **d**	0.29 ± 0.02 **cd**	4.74 ± 0.28 **a**
AcdP	5.76 ± 0.90 **bc**	3.41 ± 0.80 **c**	11.78 ± 1.70 **b**	2.87 ± 0.98 **c**	10.98 ± 0.94 **b**	18.09 ± 0.7 **a**
AlkP	0.36 ± 0.09 **c**	4.17 ± 0.70 **b**	2.79 ± 0.47 **c**	1.02 ± 0.57 **c**	12.07 ± 3.01 **a**	4.02 ± 0.52 **b**
Aryl	1.57 ± 0.27 **bc**	nd	8.42 ± 1.22 **b**	1.04 ± 0.36 **c**	1.12 ± 0.50 **bc**	56.0 ± 3.5 **a**
FDA	1.22 ± 0.12 **b**	0.89 ± 0.10 **b**	1.28 ± 0.09 **b**	0.52 ± 0.04 **c**	0.49 ± 0.02 **c**	3.22 ± 0.21 **a**

3.3. Mercury Environmental Speciation

- Total mercury

Mercury levels (Table 3) were homogeneous between the rehabilitated sites with an average of 265 ng Hg g^{-1}. The Yao3 site had significantly higher mercury content with 338.8 ng Hg g^{-1}. These concentrations were still low compared to the gold mined soil in French Guiana [23]; however, it remained within the normal range of mercury measurements in French Guiana [23]. The total mercury is significantly and positively correlated with microbial biomass, total organic carbon and total iron, and negatively correlated with clay, total aluminum, and pH (Table 4).

Table 3. Mercury environmental speciation for the six sites. E1: Extraction of highly mobile mercury species, F1: H_2O soluble mercury fraction, F2: $CaCl_2$ exchangeable mercury fraction, F3: EDTA bioavailable mercury fraction, E2: Extraction of mobile and toxic mercury species, E3: Extraction of semi mobile mercury species, E4: Non-mobile mercury species. All values are expressed in ng g^{-1} soil. Extraction yield represents the quality of mercury extraction in the samples, the value is obtained with the following formula: (E1 + E2 + E3 + E4)/Total Hg × 100. For each parameter, values followed by different letters differ significantly with $p < 0.05$ with Tukey HSD test.

Extraction	Bel1	Bel2	Bel3	Yao1	Yao2	Yao3
E1	6.28 ± 2.16 **a**	7.30 ± 0.48 **a**	6.82 ± 0.49 **a**	18.32 ± 1.29 **b**	8.90 ± 0.88 **a**	4.07 ± 0.52 **a**
F1	4.87 ± 0.09 **b**	4.97 ± 0.12 **b**	4.21 ± 0.08 **b**	10.39 ± 0.82 **a**	5.55 ± 0.45 **b**	1.71 ± 0.21 **b**
F2	0.36 ± 2.06 **a**	1.27 ± 0.47 **b**	1.33 ± 0.43 **b**	1.23 ± 0.83 **b**	1.89 ± 0.54 **b**	1.28 ± 0.28 **b**
F3	1.04 ± 0.09 **a**	1.04 ± 0.19 **a**	1.27 ± 0.29 **a**	6.69 ± 0.19 **b**	1.46 ± 0.27 **a**	1.08 ± 0.53 **a**
E2	65.8 ± 14.9 **ab**	14.4 ± 1.6 **bc**	44.1 ± 8.3 **b**	70.4 ± 37.7 **ab**	103.1 ± 5.3 **a**	25.9 ± 4.0 **b**
E3	69.2 ± 5.1 **c**	132.5 ± 5.9 **b**	167.8 ± 9.8 **a**	118.6 ± 7.1 **bc**	92.8 ± 3.6 **bc**	128.6 ± 14.9 **b**
E4	147.1 ± 12.2 **a**	69.2 ± 4.0 **ab**	53.9 ± 15.2 **ab**	10.1 ± 3.7 **b**	23.1 ± 1.6 **b**	184.0 ± 4.8 **a**
Total Hg	292.8 ± 16.5 **ab**	224.1 ± 7.2 **b**	273.9 ± 4.6 **ab**	228.5 ± 11.5 **b**	232.7 ± 3.9 **b**	338.8 ± 11.6 **a**
Extraction yield (%)	98.5	99.7	99.5	103.1	97.9	101.1

Table 4. Pearson correlation coefficients (n = 30) between enzymes activities, soil parameters and mercury mobility; bold values are significant at the $p < 0.05$.

	E1	E2	E3	E4	F1	F2	F3	HgT
MBC	**−0.90**	**−0.84**	−0.26	**0.84**	**−0.93**	**−0.60**	**−0.83**	**0.92**
DH	**−0.89**	**−0.78**	−0.37	**0.85**	**−0.93**	**−0.59**	**−0.81**	**0.93**
Bglu	**−0.89**	**−0.79**	−0.39	**0.86**	**−0.93**	**−0.59**	**−0.81**	**0.93**
Urease	**−0.90**	**−0.84**	−0.35	**0.88**	**−0.93**	**−0.62**	**−0.82**	**0.93**
P5	**−0.91**	**−0.64**	**−0.42**	**0.79**	**−0.94**	**−0.51**	**−0.87**	**0.91**
P9	0.09	**0.45**	0.14	−0.37	0.09	**0.72**	−0.05	−0.31
Aryl	**−0.88**	**−0.81**	**−0.40**	**0.88**	**−0.91**	**−0.63**	**−0.79**	**0.94**
FDA	**−0.92**	**−0.84**	**−0.45**	**0.93**	**−0.94**	**−0.72**	**−0.83**	**0.97**
Clay	**0.94**	**0.59**	**0.69**	**−0.90**	**0.92**	**0.75**	**0.94**	**−0.86**
TOC	**−0.90**	**−0.86**	−0.30	**0.87**	**−0.93**	**−0.60**	**−0.83**	**0.92**
Al	**0.68**	**0.70**	**0.43**	**−0.79**	**0.66**	**0.92**	**0.60**	**−0.75**
Fe	**−0.46**	−0.12	−0.10	0.22	**−0.48**	−0.22	**−0.44**	**0.47**
pH	**0.70**	**0.83**	−0.08	**−0.62**	**0.73**	0.20	**0.70**	**−0.58**

- Total highly mobile and mobile mercury species

The quantities of highly mobile mercury E1 (Table 3), and their percentage in relation to total mercury (Figure 2b), were significantly higher for site Yao1 with a percentage of 8% for E1. Yao1 was also the richest in water soluble mercury (F1) and bioavailable mercury (F3) with 4.56 and 2.93%, respectively. For the other rehabilitated sites, the percentages of fractions F1, F2 and F3 (Figure 2a) did not show significant differences. Concerning mobile and toxic (E2) mercury, the Bel2 and Yao3 sites had the lowest grades with percentages of 6.46% and 7.68%, while Yaoni's two non-restored sites had average grades of 31.1% and 44.3% respectively. E1 and E2 are correlated negatively with microbial parameters and TOC and positively with total clay and aluminum (Table 4).

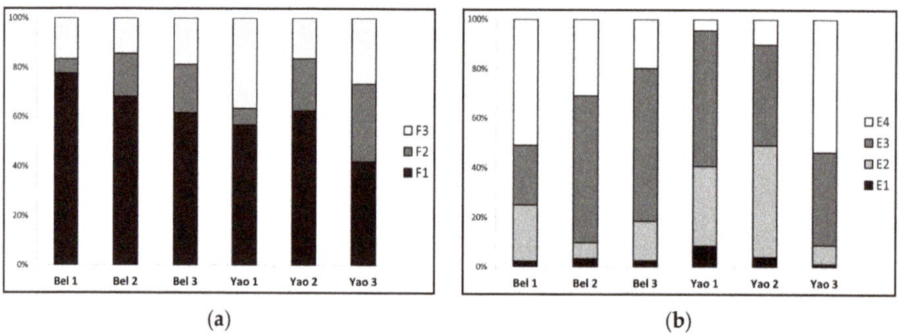

Figure 2. Mercury environmental speciation for the six sites as a percentage of each fraction relative to total mercury. (**a**) Focus on the most mobile fractions of mercury. F1: H_2O soluble mercury, F2: $CaCl_2$ exchangeable mercury, F3: EDTA bioavailable mercury. (**b**) Focus on the main extractions of mercury. E1: Extraction of highly mobile mercury, E2: Extraction of mobile and toxic mer cury, E3: Extraction of semi mobile mercury, and E4: Non-mobile mercury.

- Total semi-mobile mercury species

The highest percentages were detected for sites Yao1, Bel2 and Bel3 with values of 51.9%, 59.2% and 61.3%, respectively. The Bel1 site had the lowest quantity and percentage of semi-mobile mercury with 23.7%. The semi-mobile mercury fraction did not have a significant effect on microbial parameters and was significantly correlated with clay and aluminum content.

- Total non-mobile mercury species

The restored Yao3 site showed that mercury was most immobilized with 54% non-mobile mercury. The two non-restored Yaoni sites Yao1 and Yao2 had significantly lower levels of non-mobile mercury than the other sites with 4.4% and 9.9%, respectively. The two non-restored sites Bel1 and Bel2 had more mercury, 50.3% and 30.9%, respectively, than the restored site Bel3 with 19.7%. The correlation between non-mobile mercury and the dataset followed exactly the same pattern as the total mercury content.

3.4. Co-Inertia between Soil Parameters and Microbial Activities Dataset

The co-inertia analysis (Figure 3a) showed that soil parameters (six variables) had a highly significant correlation ($p < 0.001$) with the microbiological dataset (eight variables) with 41.86% of total variance explained. Total organic carbon was positively correlated to enzyme activities with the exception of acid and alkaline phosphatase. Soil pH, fine sand and clay content were negatively related to the microbiological dataset.

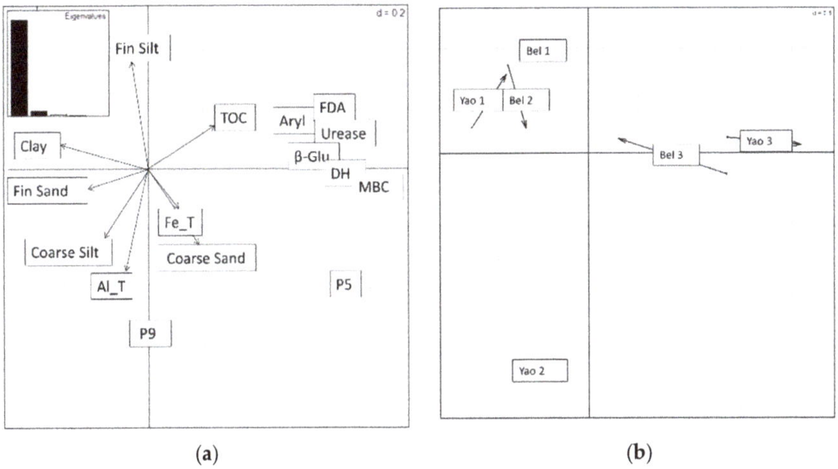

Figure 3. (**a**) Co-inertia analysis between soil physico-chemical parameters and microbiological variables. The variables linked by lines refer to the physico-chemical variables, the other variables refer to the different enzymes activities and microbial biomass in the soil sample; (**b**) discriminant analysis on co-inertia variable of soil samples. Boxes represent the projected co-ordinates of physico-chemical dataset and microbial dataset of each rehabilitated site, respectively. The length of the arrow is proportional to the divergence between the data sets. Eigen values 86.74% and 9.08% for axis 1 to 2, respectively. Randtest: simulated p-value: 0.0019. Explained variance: 41.81%.

Discriminant analysis of co-inertia (Figure 3b) showed that soil samples were significantly distinguished ($p < 0.001$) according to their spatial distribution in terms of the quality of ecological restoration. In the axis 1 direction, soil samples collected from restored sites (Bel3 and Yao3) were predominated on the positive axis, and were clearly distinguished from non-restored sites, all on the negative side.

According to these data, the microbial biomass and all enzyme activities tested except for phosphatases would be correlated exclusively to the organic carbon concentration. However, we observed a negative correlation of the microbiological data with the concentration of clays and fine sands. Texture and structure are among the most important factors influencing the activity of telluric enzymes [39]. Enzyme activity is generally greater in fine textured soils than in coarse textured soils [40,41] and negative correlations between enzyme activity and clays appear to show opposite results. However, while the nature of clays also impacts enzymatic activities [42,43], it has been shown that the structural stability of soil aggregates, linked to the organic matter content, could play a more important role than soil texture alone [40,44,45]. Organic matter stabilizes aggregate structure, increases soil retention capacity and improves nutrient bioavailability [46,47], thus stimulating soil microbial flora [48].

In this study, the major factor of microbial activity was organic matter. In these rehabilitated soils, the contribution of organic matter is exclusively conditioned by the nature of the plant cover and therefore by the success of ecological restoration. The choice of the implantation of fabaceous species, fast-growing nitrogen-fixing species, seems judicious for their ability to bring a large amount of organic matter to the soil [49,50].

3.5. Co-Inertia between Soil Parameters and Environmental Mercury Speciation

The co-inertia analysis (Figure 4a) showed that soil parameters (eight variables) had a highly significant correlation ($p < 0.001$) with the environmental mercury speciation dataset (seven variables) with 82.06% of total variance explained, which could indicate relatively similar mercury retention and mobilization mechanisms between different rehabilitated sites.

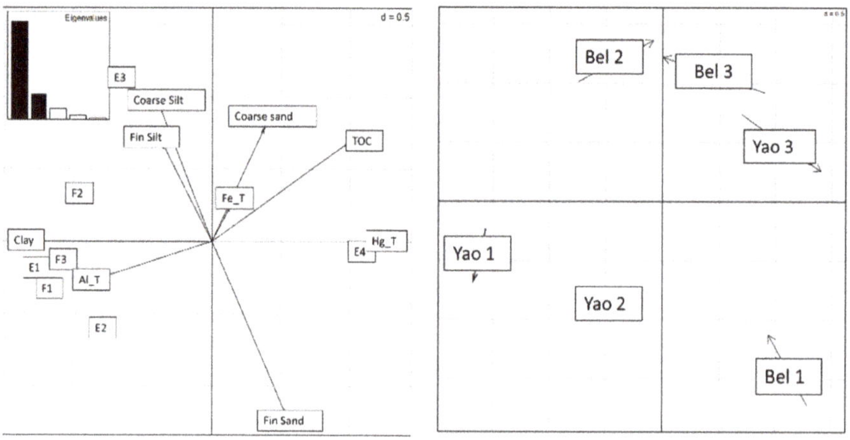

(a) (b)

Figure 4. (**a**) Co-inertia analysis between soil physico-chemical parameters and mercury extraction. The variables linked by lines refer to the physico-chemical variables, the other variables refer to the different extractions and fractions of mercury in the samples; (**b**) discriminant analysis on co-inertia variable of soil samples. Circles and boxes represent the projected co-ordinates of physico-chemical dataset and mercury dataset of each rehabilitated site, respectively. The length of the arrow is proportional to the divergence between the data sets. Eigen values 67.96% and 20.34% for axis 1 to 2, respectively. Randtest: simulated p-value: 0.0001. Explained variance: 82.06%.

Clay was strongly correlated with highly mobile mercury species (E1, F1, and F3) and negatively correlated to total and non-mobile mercury, while silt was related to exchangeable mercury and semi-mobile mercury species. Mobile mercury species were correlated with coarse sand, but not with fine sand. Soil pH was significantly and positively correlated (Table 4) with mobile mercury species (E2) and negatively correlated with total Fe content.

Discriminant analysis of co-inertia (Figure 4b) showed that soil samples were significantly distinguished ($p < 0.001$) according to their spatial distribution in terms of the quality of ecological restoration. In axis 1 direction, soil samples collected from restored sites (Bel3 and Yao3) were predominated on the positive axis and were clearly distinguished from non-restored sites. The highly mobile clay and mercury variable appeared to strongly distinguish the Yaoni sites, while the percentages of silt, sand and semi-mobile mercury appeared to more differentiate the Belizon sites along axis 2.

4. Discussion

4.1. Positive Effect of Ecological Restoration on Mercury Mobility

In this study, the results showed very distinct mercury behaviors based on sites that were not solely related to vegetation density. Indeed, the three least covered sites (Bel1, Yao1 and Yao2) presented very different speciation results with nearly 50% non-mobile mercury for Bel1 and less than 10% non-mobile mercury for Yao1 and Yao2. In addition, the results showed a significant increase in mobile and toxic mercury for sites Yao1 and Yao2 with more than 40% compared to about 25% of mobile mercury for Bel1. The presence of tree cover did not positively impact the mobility of mercury on the Bel3 site compared to adjacent plot Bel2, which showed a higher proportion of semi-extractable and non-mobile mercury and less mobile mercury. However, mercury speciation over the oldest restored area of Yao3 showed a very significant decrease in mobility and potential toxicity of mercury with 90% of mercury present in non-extractable forms.

While HCL/ethanol extraction represents more of a mobility potential linked to leaching under acidic conditions, water extraction, $CaCl_2$ and EDTA represent more direct mobility close to natural conditions in soils. The F1, F2 and F3 extractions showed the lowest percentage of easily mobile mercury for the Yao3 site and the highest for Yao1 with 1 and 8%, respectively. While the low mobility of mercury species for the Yao3 site are close to those for the Asturias [51], Almaden [52] and Usagre [53] mines in Spain, the Cuyuni Basin [54] in Venezuela, old mining mines in Mexico [55], in the Yaoni region of French Guiana [16], and the mercury mobility and toxicity for Yao1 were significantly higher and comparable to mobility at some Wanshan mining sites in China [56]. Depending on rainfall, these sites can go into suboxic or even anoxic conditions and these results show that the pool of mercury available for rapid methylation diverges very strongly between the rehabilitated sites.

The results indicate the contribution of a dense cover of fabaceae on the Yao3 site to modify mercury speciation compared to the unrestored Yao1 and Yao2 sites. At this site, despite a higher total mercury content than at the other rehabilitated sites, most of the mercury corresponds to very stable chemical species with low mobility and high stability, such as amalgamated elemental mercury, strongly bound mercury (mineralized and metal-oxyhydroxides) and species in the form of sulphides and hydroxide oxides. For the restored site Bel3, the results are more contrasted, but are possibly due to the more recent restoration and more limited contributions in organic matter in only four years of vegetation recovery.

4.2. Effect of Soil and Microbial Properties on Environmental Speciation of Mercury

In view of the above conclusions, the nature of the vegetation cover is not only indicative of mercury mobility. The mercury speciation provide access to information on toxicity, but do not provide a clear picture of the influence of mercury-carrying phases in soils. In tropical soils, the mobility and retention of mercury are largely from the nature of carrier phases. More specifically, clay minerals, iron and aluminum (oxyhydr)oxides, and the quantity and nature of organic matter represent the most important carrier phases and vary according to the biochemical conditions of the environment [57–60].

Extractions E1, E2 as well as the highly soluble fractions F1, F2 and F3, are positively correlated with clay and total aluminum contents. In tropical soils, rich in kaolinite and aluminosilicates, the adsorption of mercury in clay minerals may be important, especially when the pH is above 5 [61]. This distribution in soil of soluble fractions of mercury in clay minerals is consistent with the literature and shows that mercury adsorbed on these soil phases appears to be readily mobilizable [61].

Concerning the relations between mercury and the organic component of the soil (organic matter and biomass), it would seem that organic matter is the main carrier phase of non-mobile and total mercury (r^2 = 0.87 and 0.92), which are results consistent with the literature [62], and are already observed on restored sites in French Guiana [16]. The increase in mercury content in restored soil in relation to organic matter content could be related to the heterogeneity of mining soils, but could also be a key factor in the processes of mercury retention in the soil and in the limitation of losses through leaching and erosion of soil particles. The enrichment of organic matter levels through ecological restoration could therefore have an additional effect by potentially limiting mercury transfers. It would also appear that these non-mobile and total mercury concentrations do not have an impact on the development of soil microbial communities with positive correlations with microbial biomass and enzymatic activities, a result already shown in some mining areas with mercury loading in soil [63]. Nevertheless, the negative correlations between E2 (toxic mercury) and microbial biomass could indicate that a predominance of toxic forms of mercury, even at relatively low concentrations, could negatively affect the development of microbial communities and their enzymatic activities. Another major point is that these potentially toxic forms of mercury do not correlate with soil organic matter. In this context, microbial mineralization of organic matter would therefore be likely to release non-toxic forms of mercury, but would have a very limited impact on the transfer of toxic mercury into the soil.

This study also shows a significant correlation between total iron and total mercury. While iron oxide concentrations in soils were not measured in this study, these results confirm the importance of iron in the biogeochemistry of mercury and its retention in soil phases. This hypothesis also seems to be confirmed by the important correlations observed between semi-mobile mercury extraction and silty soil fractions, rich in metal-oxyhydroxides such as goethite or hematite. This study showed that the impact of non-homogeneous mine site remediation processes has altered mercury speciation by acting directly on soil texture. These results also showed that the level of vegetation cover, and the level of organic matter brought to the soil, has improved mercury retention and sanitary quality.

4.3. *Positive Effect of Ecological Restoration on Microbial Activities*

Soil biochemical parameters such as microbial biomass or soil enzymatic activities are sensitive and relevant indicators of stressed and disturbed soils and are used to describe soil quality status [15,64]. Generally, microbial biomass varies according to soil physico-chemical parameters, vegetation cover or potential contamination [15,65]. In this study, microbial biomass appeared to be very sensitive to the different modalities of ecological restoration and to the nature of the vegetation cover. The presence of a restored tree cover on the Bel3 and Yao3 plots had increased the microbial biomass by a factor of 2–4 compared to non-restored sites. The presence of *Lycopodiella* sp. on the Bel2 plot has also positively stimulated microbial biomass and shows that the increase in plant density, even herbaceous, has a significant impact on microbial biomass.

Soil enzymatic activities represent the overall degradation activity of organic soil components and are affected by many factors, such as pH, organic matter content or contamination level [66], and are mediated by microbial biomass. In our study, the restored Bel3 and Yao3 sites had the highest activities, except for alkaline phosphatase activities. However, the classification of soils according to the amount of microbial biomass is not the same as for enzymatic activities. Indeed, the Yao2 site, which has one of the lowest biomasses, has relatively important dehydrogenase, B-glucosidase, acid and alkaline phosphatase activities. This could show that these enzymatic activities are not only related to microbial biomass. These results indicated a significant and positive effect of controlled ecological restoration on certain components of biochemical activities in rehabilitated soils [67–70]. The very low level of activity on non-restored soils also highlights the stressed and anthropized nature of these former mining lands, which require adapted ecological rehabilitation protocols.

4.4. *Footprint of Rehabilitation on Soil Functionning and Mercury Mobility*

The various stages of rebuilding an ecosystem after human degradation, such as mining are complex, multidisciplinary and often involve ecological engineering. On mining sites in French Guiana, these protocols have to be adapted according to the degree of site contamination, soil texture, water regime and of course the nature of the plant species used for restoration [14]. *Clitoria racemosa* is a fabacea native to South America and favors sandy-gravelly or even sandy-clay substrates. At the Belizon mine, difficult site rehabilitation and earthworks led to a high degree of textural heterogeneity, resulting in a mosaic of vegetation and partial success in ecological restoration. The recovery of the macrocuts was not effective on the overly draining and sandy substrate of the Bel1 plots, now colonized by *Cyperaceae* sp. On the Bel2 plots, the significant presence of silt led to a rapid covering of the site by *Lycopodiella* sp., preventing any recovery of *C. racemosa* plants. The rational use of other products, such as *Acacia mangium*, adapted to gravel surfaces, and *Erythrina fusca*, adapted to silty substrates on the Bel1 and Bel2 plots, respectively, could have allowed for a more widespread success of ecological restoration. These results showed the importance of adapting ecological restoration protocols according to the nature of the substrate, as already described by Loubry [14].

On the former Yaoni mine, the very significant differences in terms of organic constituents of the soil provide interesting elements for the restoration of mining sites. The very low CNP stocks on plots Yao1 and Yao2 showed that mining activity depletes organic constituents and soil fertility. Seventeen years after the end of the rehabilitation, these sites had still not found shrub or tree cover and ecological succession seemed to be stopped. The controlled recovery of vegetation on the Yao3 site has highlighted the positive effect of the establishment of a cover of fabaceae on the renewal of the stock of nutrients and organic matter. It would appear that the organic matter brought in during ecological restoration is also a driving force behind the retention of mercury and the limitations of potential transfers to aqueous media. A preliminary study [16] showed that ecological restoration could affect the nature of the mercury bearing phases. However, it would also seem that the potential toxicity of these plots under ecological restoration could be reduced by the establishment of a plant cover.

5. Conclusions

We evaluated the effect of the rehabilitation of Guyana's mining sites in order to link soil texture with the quality of ecological restoration to the biochemical quality of the soil and the potential toxicity of mercury. The results indicated very variable and heterogeneous soil textures, even on nearby sites. Some sites rehabilitated with too much silt or sand have not allowed for the ecological restoration of a tree canopy. The results also highlight a direct impact of the success of ecological restoration on the return of microbial activities related to biogeochemical cycles. The provision of bedding using fabaceous species seems to be one of the key factors for the success of ecological restoration. The results also showed an impact of soil texture on mercury mobility and toxicity. Environmental speciation measures showed a link between mercury mobility and the nature of rehabilitation. The least restored sites have potentially higher concentrations of mobile and semi-mobile mercury, related to the quantities of clays and silts. The homogeneous texture of the restored Yaoni site, as well as the establishment of a dense tree canopy, has contributed to limiting mercury mobility by modifying its speciation towards more insoluble forms, such as cinnabar.

Author Contributions: Conceptualization: N.B. and E.C., The sampling campaigns for the sites presented in this study are the result of a second sampling campaign on mining sites undergoing ecological restoration. The samples were taken by E.C. and N.B., with Christian Pernaut as field guide. The methodology: E.C. and N.B.; validation, N.B.; formal analysis, E.C., V.A., A.L. and S.G.-M.; investigation, E.C.; resources, E.C.; data curation, E.C.; writing—original draft preparation, E.C.; writing—review and editing, E.C.; visualization, E.C.; supervision, N.B.; project administration, N.B.; funding acquisition, N.B. All authors have read and agreed to the published version of the manuscript.

Funding: This research received no external funding.

Institutional Review Board Statement: Not applicable.

Informed Consent Statement: Not applicable.

Acknowledgments: We would like to thank Christian Pernaut, the manager of the mining company Ermina (Matoury-France), who was our guide in French Guiana during the sampling campaign. We thank Philippe Matheus, the manager of the mining company C. M. B (Remire Montjoly-France). for his logistical and material support. We also thank the Cayenne and Bondy Development Research Institute for their analytical contributions.

Conflicts of Interest: The authors declare no conflict of interest.

References

1. Van Der Werf, G.R.; Morton, D.C.; Defries, R.S.; Olivier, J.G.J.; Kasibhatla, P.S.; Jackson, R.B.; Collatz, G.J.; Randerson, J.T. CO_2 emissions from forest loss. *Nat. Geosci.* **2009**. [CrossRef]
2. Achard, F.; Beuchle, R.; Mayaux, P.; Stibig, H.J.; Bodart, C.; Brink, A.; Carboni, S.; Desclée, B.; Donnay, F.; Eva, H.D.; et al Determination of tropical deforestation rates and related carbon losses from 1990 to 2010. *Glob. Chang. Biol.* **2014**. [CrossRef] [PubMed]
3. Song, X.P.; Hansen, M.C.; Stehman, S.V.; Potapov, P.V.; Tyukavina, A.; Vermote, E.F.; Townshend, J.R. Global land change from 1982 to 2016. *Nature* **2018**. [CrossRef]
4. Veiga, M.M.; Maxson, P.A.; Hylander, L.D. Origin and consumption of mercury in small-scale gold mining. *J. Clean. Prod.* **2006**, *14*, 436–447. [CrossRef]
5. Wong, M. Ecological restoration of mine degraded soils, with emphasis on metal contaminated soils. *Chemosphere* **2003**, *50*, 775–780. [CrossRef]
6. Grimaldi, M.; Guédron, S.; Grimaldi, C. Impact of gold mining on mercury contamination and soil degradation in Amazonian ecosystems of French Guiana. In *Land-Use Change Impacts on Soil Processes: Tropical and Savannah Ecosystems*; CABI International: Wallingford, CT, USA, 2015; pp. 95–107, ISBN 9781780642109.
7. Telmer, K.; Costa, M.; Angélica, R.S.; Araujo, E.S.; Maurice, Y. The source and fate of sediment and mercury in the Tapajós River, Pará, Brazilian Amazon: Ground- and space-based evidence. *J. Environ. Manag.* **2006**, *81*, 101–113. [CrossRef]
8. Roulet, M.; Lucotte, M.; Farella, N.; Serique, G.; Coelho, H.; Passos, C.S.; Da Silva, E.D.; De Andrade, P.S.; Mergler, D.; Guimarães, J.R.; et al. Effects of Recent Human Colonization on the Presence of Mercury in Amazonian Ecosystems. *Water Air. Soil Pollut.* **1999**, *112*, 297–313. [CrossRef]
9. Wasserman, J.C.; Fundação, S.S.H.; Cruz, O.; Wasserman, M. Biogeochemistry of Mercury in the Amazonian Environment Mercury Exposure in Yanomami Indigenous Communities in the sate of Roraima View project Fate of nanoparticles in the environment View project. *Ambio J. Hum. Environ.* **2003**, *32*, 336–342. [CrossRef]
10. Lacerda, L.D.; De Souza, M.; Ribeiro, M.G. The effects of land use change on mercury distribution in soils of Alta Floresta, Southern Amazon. *Environ. Pollut.* **2004**, *129*, 247–255. [CrossRef] [PubMed]
11. Guedron, S. Impact de l 'Exploitation Minière en Guyane Française Sur Les Flux de Mercure Vers Les Écosystèmes Aquatiques. Ph.D. Thesis, Université Joseph Fourier—Grenoble I, Grenoble, France, 14 March 2008.
12. Alloway, B. *Heavy Metals in Soils:Trace Metals and Metalloids in Soils and Their Bioavailability*; Springer: Dordrecht, The Netherlands, 2012; ISBN 9789400744691.
13. Veiga, M.M.; Baker, R.F. *Protocols for Environmental and Health Assessment of Mercury Released by Artisanal and Small-Scale Gold Miners*; United Nations: New York, NY, USA, 2004; ISBN 9211064295.
14. Han, Y.; Kingston, H.M.; Boylan, H.M.; Rahman, G.M.M.; Shah, S.; Richter, R.C.; Link, D.D.; Bhandari, S. Speciation of mercury in soil and sediment by selective solvent and acid extraction. *Anal. Bioanal. Chem.* **2003**, *375*, 428–436. [CrossRef]
15. Navarro, A. Review of characteristics of mercury speciation and mobility from areas of mercury mining in semi-arid environments. *Rev. Environ. Sci. Biotechnol.* **2008**, *7*, 287–306. [CrossRef]
16. Basavaraja, P.K.; Sharma, S.D.; Dhananjaya, B.N.; Badrinath, M.S. Acacia nilotica: A tree species for amelioration of sodic soils in Central dry zone of Karnataka, India. In Proceedings of the 19th World Congress of Soil Science, Soil Solutions for a Changing World, Brisbane, Australia, 1–6 August 2010; pp. 73–76.
17. Balland-Bolou-Bi, C.; Turc, B.; Alphonse, V.; Bousserrhine, N. Impact of microbial communities from tropical soils on the mobilization of trace metals during dissolution of cinnabar ore. *J. Environ. Sci.* **2017**, *56*, 122–130. [CrossRef] [PubMed]
18. Le Roux, C. La réhabilitation des mines et carrières à ciel ouvert: Restauration des sites miniers. *Bois Forêts Trop.* **2002**, *272*, 5–19.
19. Jaffré, T.; McCoy, S.; Rigault, F.; Dagostini, G. Quelle méthode de végétalisation pour la réhabilitation des anciens sites miniers de Nouvelle-Calédonie. In *Écologie des milieux sur roches ultramafiques sur sols métallifères actes la deuxième conférence Int. sur l'écologie des milieux serpentiniques*; ORSTOM: Noumea, France, 1997; pp. 285–288.
20. Loubry, D. *Livret Technique Pour la Conduite de la Revégétalisation sur les Surfaces Minières Alluvionanaires de Guyane*; Unpublished Technical Report; IRD: Cayenne, Guyane, 2002; 51p.
21. Nannipieri, P.; Ascher, J.; Ceccherini, M.T.; Landi, L.; Pietramellara, G.; Renella, G. Microbial diversity and soil functions. *Eur. J. Soil Sci.* **2003**, *54*, 655–670. [CrossRef]
22. Couic, E.; Grimaldi, M.; Alphonse, V.; Balland-Bolou-Bi, C.; Livet, A.; Giusti-Miller, S.; Sarrazin, M.; Bousserrhine, N. Mercury behaviour and C, N, and P biogeochemical cycles during ecological restoration processes of old mining sites in French Guiana. *Environ. Sci. Process. Impacts* **2018**, *20*. [CrossRef]

23. Guedron, S.; Grangeon, S.; Lanson, B.; Grimaldi, M. Mercury speciation in a tropical soil association; Consequence of gold mining on Hg distribution in French Guiana. *Geoderma* **2009**, *153*, 331–346. [CrossRef]
24. Schimann, H.; Petit-Jean, C.; Guitet, S.; Reis, T.; Domenach, A.M.; Roggy, J.C. Microbial bioindicators of soil functioning after disturbance: The case of gold mining in tropical rainforests of French Guiana. *Ecol. Indic.* **2012**, *20*, 34–41. [CrossRef]
25. Leermakers, M.; Baeyens, W.; Quevauviller, P.; Horvat, M. Mercury in environmental samples: Speciation, artifacts and validation. *TrAC Trends Anal. Chem.* **2005**, *24*, 383–393. [CrossRef]
26. Murphy, J.; Riley, J.P. A modified single solution method for the determination of phosphate in natural waters. *Anal. Chim. Acta* **1962**, *27*, 31–36. [CrossRef]
27. Anderson, J.P.E.; Domsch, K.H. A physiological method for the quantitative measurement of microbial biomass in soils. *Soil Biol. Biochem.* **1978**, *10*, 215–221. [CrossRef]
28. Badiane, N.N.Y.; Chotte, J.L.; Pate, E.; Masse, D.; Rouland, C. Use of soil enzyme activities to monitor soil quality in natural and improved fallows in semi-arid tropical regions. *Appl. Soil Ecol.* **2001**, *18*, 229–238. [CrossRef]
29. Mora, P.; Miambi, E.; Jiménez, J.J.; Decaëns, T.; Rouland, C. Functional complement of biogenic structures produced by earthworms, termites and ants in the neotropical savannas. *Soil Biol. Biochem.* **2005**, *37*, 1043–1048. [CrossRef]
30. Tabatabai, M.A.; Bremner, J.M. Assay of urease activity in soils. *Soil Biol. Biochem.* **1972**, *4*, 479–487. [CrossRef]
31. Klein, D.A.; Loh, T.C.; Goulding, R.L. A rapid procedure to evaluate the dehydrogenase activity of soils low in organic matter. *Soil Biol. Biochem.* **1971**, *3*, 385–387. [CrossRef]
32. Kandeler, E.; Gerber, H. Short-term assay of soil urease activity using colorimetric determination of ammonium. *Biol. Fertil. Soils* **1988**, *6*, 68–72. [CrossRef]
33. Green, V.S.; Stott, D.E.; Diack, M. Assay for fluorescein diacetate hydrolytic activity: Optimization for soil samples. *Soil Biol. Biochem.* **2006**, *38*, 693–701. [CrossRef]
34. Rasmussen, L.D.; Sørensen, S.J.; Turner, R.R.; Barkay, T. Application of a mer-lux biosensor for estimating bioavailable mercury in soil. *Soil Biol. Biochem.* **2000**, *32*, 639–646. [CrossRef]
35. Silveira, M.L.; Alleoni, L.R.F.; O'Connor, G.A.; Chang, A.C. Heavy metal sequential extraction methods—A modification for tropical soils. *Chemosphere* **2006**, *64*, 1929–1938. [CrossRef] [PubMed]
36. Harris-Hellal, J.; Vallaeys, T.; Garnier-Zarli, E.; Bousserrhine, N. Effects of mercury on soil microbial communities in tropical soils of French Guyana. *Appl. Soil Ecol.* **2009**, *41*, 59–68. [CrossRef]
37. Harris-Hellal, J.; Grimaldi, M.; Garnier-Zarli, E.; Bousserrhine, N. Mercury mobilization by chemical and microbial iron oxide reduction in soils of French Guyana. *Biogeochemistry* **2011**, *103*, 223–234. [CrossRef]
38. Da Silva, E.; Nahmani, J.; Lapied, E.; Alphonse, V.; Garnier-Zarli, E.; Bousserrhine, N. Toxicity of mercury to the earthworm Pontoscolex corethrurus in a tropical soil of French Guiana. *Appl. Soil Ecol.* **2016**, *104*, 79–84. [CrossRef]
39. Shukla, G.; Varma, A. *Soil Enzymology*; Springer: Berlin/Heidelberg, Germany, 2011; ISBN 9783642142246.
40. Taylor, J.P.; Wilson, B.; Mills, M.S.; Burns, R.G. Comparison of microbial numbers and enzymatic activities in surface soils and subsoils using various techniques. *Soil Biol. Biochem.* **2002**, *34*, 387–401. [CrossRef]
41. Acosta-Martínez, V.; Cruz, L.; Sotomayor-Ramírez, D.; Pérez-Alegría, L. Enzyme activities as affected by soil properties and land use in a tropical watershed. *Appl. Soil Ecol.* **2007**. [CrossRef]
42. Huang, Q.; Liang, W.; Cai, P. Adsorption, desorption and activities of acid phosphatase on various colloidal particles from an Ultisol. *Colloids Surf. B Biointerfaces* **2005**. [CrossRef]
43. Safari Sinegani, A.A.; Emtiazi, G.; Shariatmadari, H. Sorption and immobilization of cellulase on silicate clay minerals. *J. Colloid Interface Sci.* **2005**. [CrossRef]
44. Caravaca, F.; Hernández, T.; García, C.; Roldán, A. Improvement of rhizosphere aggregate stability of afforested semiarid plant species subjected to mycorrhizal inoculation and compost addition. *Geoderma* **2002**, *108*, 133–144. [CrossRef]
45. Udawatta, R.P.; Kremer, R.J.; Adamson, B.W.; Anderson, S.H. Variations in soil aggregate stability and enzyme activities in a temperate agroforestry practice. *Appl. Soil Ecol.* **2008**. [CrossRef]
46. Hobbs, R.J.; Harris, J.A. Restoration ecology: Repairing the earth's ecosystems in the new millennium. *Restor. Ecol.* **2001**, *9*, 239–246. [CrossRef]
47. Viana, R.M.; Ferraz, J.B.S.; Neves, A.F.; Vieira, G.; Pereira, B.F.F. Soil quality indicators for different restoration stages on Amazon rainforest. *Soil Tillage Res.* **2014**, *140*, 1–7. [CrossRef]
48. Elfstrand, S.; Båth, B.; Mårtensson, A. Influence of various forms of green manure amendment on soil microbial community composition, enzyme activity and nutrient levels in leek. *Appl. Soil Ecol.* **2007**, *36*, 70–82. [CrossRef]
49. Butterfield, R.P. Early species selection for tropical reforestation: A consideration of stability. *For. Ecol. Manag.* **1996**, *81*, 161–168. [CrossRef]
50. Norisada, M.; Hitsuma, G.; Kuroda, K.; Yamanoshita, T.; Masumori, M.; Tange, T.; Yagi, H.; Nuyim, T.; Sasaki, S.; Kojima, K. Acacia mangium, a nurse tree candidate for reforestation on degraded sandy soils in the Malay Peninsula. *For. Sci.* **2005**, *51*, 498–510.
51. Fernández-Martínez, R.; Larios, R.; Gómez-Pinilla, I.; Gómez-Mancebo, B.; López-Andrés, S.; Loredo, J.; Ordóñez, A.; Rucandio, I. Mercury accumulation and speciation in plants and soils from abandoned cinnabar mines. *Geoderma* **2015**, *253–254*, 30–38. [CrossRef]

52. García-Sánchez, A.; Murciego, A.; Álvarez-Ayuso, E.; Regina, I.S.; Rodríguez-González, M.A. Mercury in soils and plants in an abandoned cinnabar mining area (SW Spain). *J. Hazard. Mater.* **2009**, *168*, 1319–1324. [CrossRef]
53. Moreno-Jiménez, E.; Gamarra, R.; Carpena-Ruiz, R.O.; Millán, R.; Peñalosa, J.M.; Esteban, E. Mercury bioaccumulation and phytotoxicity in two wild plant species of Almadén area. *Chemosphere* **2006**. [CrossRef]
54. Santos-Francés, F.; García-Sánchez, A.; Alonso-Rojo, P.; Contreras, F.; Adams, M. Distribution and mobility of mercury in soils of a gold mining region, Cuyuni river basin, Venezuela. *J. Environ. Manag.* **2011**. [CrossRef]
55. Gavilán-García, I.; Santos-Santos, E.; Tovar-Gálvez, L.R.; Gavilán-García, A.; Suárez, S.; Olmos, J.; La, B.; Ticomán, L.; Delegación, G.A.; Madero, M.D.F. Mercury Speciation in Contaminated Soils from Old Mining Activities in Mexico Using a Chemical Selective Extraction. *J. Mex. Chem. Soc.* **2008**, *52*, 263–271. [CrossRef]
56. Lin, Y.; Larssen, T.; Vogt, R.D.; Feng, X. Identification of fractions of mercury in water, soil and sediment from a typical Hg mining area in Wanshan, Guizhou province, China. *Appl. Geochem.* **2010**. [CrossRef]
57. Do Valle, C.M.; Santana, G.P.; Augusti, R.; Egreja Filho, F.B.; Windm??ller, C.C. Speciation and quantification of mercury in Oxisol, Ultisol, and Spodosol from Amazon (Manaus, Brazil). *Chemosphere* **2005**, *58*, 779–792. [CrossRef]
58. Roulet, M.; Lucotte, M. Geochemistry of mercury in pristine and flooded ferralitic soils of a tropical rain forest in French Guiana, South America. *Water Air Soil Pollut.* **1995**, *80*, 1079–1088. [CrossRef]
59. Skyllberg, U.; Bloom, P.R.; Qian, J.; Lin, C.M.; Bleam, W.F. Complexation of mercury(II) in soil organic matter: EXAFS evidence for linear two-coordination with reduced sulfur groups. *Environ. Sci. Technol.* **2006**, *40*, 4174–4180. [CrossRef] [PubMed]
60. Alloway, B.J. (Ed.) *Heavy Metals in Soils*; Environmental Pollution; Springer: Dordrecht, The Netherlands, 2013; Volume 22, ISBN 978-94-007-4469-1.
61. Anderson, A. The Biochemistry of Mercury in the Environment. In *Mercury in Soils*; Elsevier/North-Holland Biomedical Press: Amsterdam, The Netherlands, 1979; pp. 79–112, ISBN 0444801103.
62. Schuster, E. The behavior of mercury in the soil with special emphasis on complexation and adsorption processes—A review of the literature. *Water Air Soil Pollut.* **1991**, *56*, 667–680. [CrossRef]
63. Campos, J.A.; Esbrí, J.M.; Madrid, M.M.; Naharro, R.; Peco, J.; García-Noguero, E.M.; Amorós, J.A.; Moreno, M.M.; Higueras, P. Does mercury presence in soils promote their microbial activity? The Almadenejos case (Almadén mercury mining district, Spain). *Chemosphere* **2018**, *201*, 799–806. [CrossRef]
64. Jordan, D.; Kremer, R.J.; Bergfield, W.A.; Kim, K.Y.; Cacnio, V.N. Evaluation of microbial methods as potential indicators of soil quality in historical agricultural fields. *Biol. Fertil. Soils* **1995**, *19*, 297–302. [CrossRef]
65. Chaussod, R. La qualité biologique des sols: Évaluation et implications. *Etude Gest. des Sols* **1996**, *3*, 261–278.
66. Niemeyer, J.C.; Lolata, G.B.; de Carvalho, G.M.; Da Silva, E.M.; Sousa, J.P.; Nogueira, M.A. Microbial indicators of soil health as tools for ecological risk assessment of a metal contaminated site in Brazil. *Appl. Soil Ecol.* **2012**. [CrossRef]
67. Hu, Y.F.; Peng, J.J.; Yuan, S.; Shu, X.Y.; Jiang, S.L.; Pu, Q.; Ma, K.Y.; Yuan, C.M.; Chen, G.D.; Xiao, H.H. Influence of ecological restoration on vegetation and soil microbiological properties in Alpine-cold semi-humid desertified land. *Ecol. Eng.* **2016**, *94*, 88–94. [CrossRef]
68. Yin, R.; Gu, C.; Feng, X.; Hurley, J.P.; Krabbenhoft, D.P.; Lepak, R.F.; Zhu, W.; Zheng, L.; Hu, T. Distribution and geochemical speciation of soil mercury in Wanshan Hg mine: Effects of cultivation. *Geoderma* **2016**, *272*, 32–38. [CrossRef]
69. Ma, Z.; Zhang, M.; Xiao, R.; Cui, Y.; Yu, F. Changes in soil microbial biomass and community composition in coastal wetlands affected by restoration projects in a Chinese delta. *Geoderma* **2017**, *289*, 124–134. [CrossRef]
70. Balland-Bolou-Bi, C.; Bolou-Bi, E.B.; Alphonse, V.; Giusti-Miller, S.; Jusselme, M.D.; Livet, A.; Grimaldi, M.; Bousserhine, N. Impact of microbial activity on the mobility of metallic elements (Fe, Al and Hg) in tropical soils. *Geoderma* **2019**, *334*, 146–154. [CrossRef]

MDPI
St. Alban-Anlage 66
4052 Basel
Switzerland
Tel. +41 61 683 77 34
Fax +41 61 302 89 18
www.mdpi.com

Applied Sciences Editorial Office
E-mail: applsci@mdpi.com
www.mdpi.com/journal/applsci

www.ingramcontent.com/pod-product-compliance
Lightning Source LLC
Chambersburg PA
CBHW062046220926
51850CB00010B/2677

* 9 7 8 3 0 3 6 5 2 1 5 3 4 *